TESTS-ANSWERS FOR FCC GENERAL RADIOTELEPHONE OPERATOR LICENSE

Revised 20th Edition

by

Warren Weagant

FCC LICENSE TRAINING

P.O. Box 3000 • Sausalito, CA 94966 • (415) 332-3161 • www.LicenseTraining.com

REVISED TWENTIETH EDITION

Library of Congress Cataloging in Publication Data Main Entry Under Title:

TESTS-ANSWERS FOR FCC GENERAL
RADIOTELEPHONE OPERATOR LICENSE

Printed in the United States of America

International Standard Book Number (ISBN): 978-0-933132-20-7

CONTENTS

The United States of America

Federal Communications Commission

GENERAL RADIOTELEPHONE OPERATOR LICENSE

(General Radiotelephone Certificate)

This certifies that the individual named below is a licensed radio operator and is authorized to operate licensed radio stations for which this class of license is valid. The authority granted is subject to any endorsement placed on this license. The authority granted is also subject to the orders, rules, and regulations of the Federal Communications Commission, the statutes of the United States, and the provisions of any treaties to which the United States is a party, which are binding upon radio operators.

This license may not be assigned or transferred to any other person. This license is valid for the lifetime of the holder unless suspended by the FCC.

Endorsement. SHIP RADAR ENDORSEMENT

Licensee: WARREN WEAGANT

Place of Issuance
SAN FRANCISCO, CA

Date of Birth	Issuance Date	License Number
JANUARY 3, 1946	JANUARY 2, 1985	PG 12 25655

Warren Weagant

Signature of Licensee

Introduction

The purpose of this testing manual is to prepare you to pass the Federal Communication's General Radiotelephone Operator License examinations. This guidebook contains the entire set of questions asked by the FCC, and provides an excellent means of preparation for the federal license exam.

A General Radiotelephone Operator License must be held by any person who adjusts, maintains, or internally repairs a radiotelephone transmitter at any station licensed by the FCC in the aviation, maritime, or international fixed public radio services.

This license must also be held by the operator of certain aviation and maritime land radio stations, compulsory equipped ship radiotelephone stations, and voluntarily equipped ship stations. Also, portable transceivers used to communicate with ships and aircraft on marine and aviation frequencies and more.

You need the Commercial General Radiotelephone License if you maintain or operate: Radio equipment on a ship that carries more than six people for hire, Transmitters on medium or high frequencies from 300 KHz to 30 MHz., Coast stations at medium or high frequency with more than 1500 watts of peak envelope power, Transmitters on ships larger than 300 gross tons where it is legally required to operate a radio station for safety purposes, International Fixed Public Radiotelelphone Stations, Certain Aircraft, Civil Air Patrol and Coast Ship Stations and repair ship radio stations, Coast transmitters, hand-held two-way radiotelephone portable radios and much more.

To obtain a Commercial FCC General Radiotelephone you must pass FCC exam Element 1 and 3 (Element 2 was discontinued). Element 1 contains Rules and Regulations questions. Element 3 contains technical and operations questions. You need to correctly answer 75% of the multiple-choice questions on the Federal exam to get your FCC License.

To obtain the FCC RADAR Endorsement for your General Radiotelephone license, you must pass the FCC Element 8 exam. This endorsement is not necessary for every license applicant. It's only for those people entering the ship-radar and avionics installation and maintenance radio services. The RADAR Endorsement is only for people who already have passed the General Radiotelephone license exam.

Examinations are no longer given at FCC Field Offices. The FCC has delegated authority to a number of private professional technical organizations (see page 12).

When you are ready to take the FCC License exam you will need to contact one of these organizations to schedule an appointment. Since each organization has a different schedule of examination dates for each area of the country - you may want to phone, email or write to several organizations to determine the best testing date and location that fits your needs.

There is no shortage of text books on electronic theory on the market. Many license applicants attempt to learn all facets of electronics while preparing for the federal exam. However, since the FCC exam covers only certain specific areas of electronic knowledge, you will save hours of study time by using this manual as your learning outline. When you discover any question in this manual you cannot answer or understand then look up the material in any one of many excellent reference books and courses containing detailed discussion of essential electronic theory.

This testing manual concentrates on specific areas covered on the federal exam. An all encompassing, general background is necessary for actual station operation and equipment maintenance, but specialized knowledge is required to successfully pass the FCC exam. If the material in this manual is completely learned and understood, you should easily answer questions on the FCC examination. It is important to note that the questions in this guide are the exact words of the questions on the actual FCC exam, although the multiple-choice answers may appear in a different order on the FCC exam.

Finally, you will notice that the format of this manual is concise and comprehensive. Everything included is relevant to the actual FCC exam. Great effort has been made to eliminate all non related material used by other publishers to "fatten out" their study guides. This is to point out that it is important to learn all test material in this training manual!

Studying on your own for the FCC examination is no easy task. It requires dedication and a great desire to operate and service sophisticated federally licensed transmitters and receivers. What's more, even when a license is not legally required for employment, this impressive license serves as a U.S. Government document certifying and acknowledging your high level of knowledge and skills in the telecommunications industry. The career advancement and additional money you will earn, the security and the attractive employment opportunities are certainly worth the effort.

Warren Weagant
Sausalito, California

Operator Licenses

A General Radiotelephone Operator License is required for anyone responsible for internal repairs, maintenance, and adjustment of FCC Licensed transmitters in the Aviation, Marine and International Public Fixed radio services.

To be eligible for this license, you must:

1. Be a legal resident - eligible for employment in the United States, including all U.S. Citizens, U.S. Nationals, and citizens of U.S. trust territories.

2. Be able to receive and transmit spoken messages in the English language.

3. Pass a multiple-choice examination covering basic radio law, operating procedures and basic electronic theory.

There are no education, training or experience requirements to obtain an operator license. Knowledge of the telegraphic code is not necessary. Your age is not important. Anyone, regardless of age, can apply for the license. This license is valid for the lifetime of the operator.

The introduction to this study guide briefly listed the authority of a person holding an FCC Commercial General Radiotelephone Operator License. Here is a more complete description of both Operating and Equipment Maintenance authority:

You need a commercial radio operator license to OPERATE the following:

SHIP Radio Stations if:

- the vessel carries more than six passengers for hire; or
- the radio operates on medium frequencies (MF) or high frequencies (HF); or
- the ship sails to foreign ports; or
- the ship station transmits radiotelegraphy; or
- the ship is larger than 300 gross tons and is required by law to carry a radio station for safety purposes.

COAST Radio Stations if:

- the radio operates on medium frequencies (MF) or high frequencies (HF); or
- the radio operates with more than 1,500 watts of peak envelope power; or
- the coast station transmits radiotelegraphy.
- Aircraft radio stations, except those which operate only on very high frequencies (VHF) and do not make foreign flights.
- International fixed public radiotelephone and radiotelegraph stations.

You need a commercial operator license to REPAIR AND MAINTAIN the following:

- All ship radio and RADAR stations.
- All coast stations.
- All hand-carried units used to communicate with ships and coast stations on marine frequencies.
- All aircraft stations and aeronautical ground stations (including hand-carried portable units) used to communicate with aircraft.
- International fixed public radiotelephone and radiotelegraph stations.

Plus, even when an FCC License is not legally required by US Federal Law for employment, it is often required by many employers in the Wireless Telecommunications, Broadcasting and many other industries.

FCC Examinations

Element 1 examination: Rules and Regulations. This is the preliminary basic requirement for higher class FCC licenses. The test covers provisions of laws, treaties and regulations with which every radio operator should be familiar, plus Radio operating procedures and practices generally followed or required in communicating by radiotelephone.

From a pool of 170 possible questions, a test administrator will select 24 questions for your exam. You must correctly answer 75% - a total of 18 questions before you pass the Element 1 exam (or an Element 3 exam test will usually not be graded).

Passing the FCC Element 3 exam certifies that you will be certified that you legally possess the operational and technical qualifications necessary to properly perform the duties required of a person holding a General Radiotelephone Operator License (GROL). To pass, you must answer correctly at least 75 out of 100 from a question pool of 600 possibilities.

Element 8 Examination (RADAR Endorsement) - Specialized theory and practice applicable to proper installation, servicing, and maintenance of radar equipment. This endorsement is not required for most license applicants - only for those persons working with RADAR navigation equipment. The FCC exam contains 50 questions from a group of 300 choices. The passing grade is 38 questions answered correctly.

All multiple-choice questions the FCC could ask you on their exam are in this testing manual, including the letters A, B, C or D designating the correct answer. The correct answer letter is printed in the column to the far right of each question:

Element 1: 144 questions in 24 Key Topic sections
Element 3: 600 questions in 100 Key Topic sections
Element 8: 300 questions in 50 Key Topic sections

The FCC has divided questions into special "Key Topics" which contain exactly 6 questions each. When you study the material in this testing manual, be aware that you will be asked one question out of every six you see in any Key Topic.

So, when you learn the correct answer to all six questions in a single topic (which is most often found on a single page in this guide), you know that only one of those questions will be on the actual FCC examination. You'll see how you are progressing with your study as you learn the answer to every question in a single "Key Topic".

How to Apply to Take the FCC Exam

For many years in the past, the FCC administered license exams at government FCC Field Offices. To prevent the staff from being overworked, the FCC conducted exams on a twice per year basis, usually during a one week period. This caused many problems for license applicants, especially if a exam was not successfully passed, forcing some people to wait for up to six months to take the exam again.

In September 1993, the FCC approved the testing for all Commercial Radiotelephone Operator License exams by the private sector. It authorized certain national private professional associations and commercial testing organizations to be classified as "Commercial Operator Licensing Examination Manager" - called "COLEM".

The use of these private organizations has made applying for and taking FCC license exams easier than ever before. Now it is possible to take the FCC examination in all 50 States, U.S. Territories, on military bases and other government sanctioned facilities around the world.

When you want to take your FCC examination, simply contact the national headquarters of a COLEM to request an examination appointment. A listing of recommended national COLEMS is included in this guidebook.

It is the responsibility of the COLEM organization to schedule exam sessions, select the specified number of questions for your exam, verify your identity, grade your examination, certify that your license exam was successfully passed and electronically file the information with the FCC Wireless Telecommunications Bureau.

Also, you should be provided an official PPC "Proof-of-Passing Certificate" for your file to keep until your actual FCC License is mailed to you by the Federal Communications Commission.

There is no limit to the number of times you may take the FCC exam. Each private testing organization may have a different waiting period between examination testing dates. However, you now have the opportunity to take the FCC exam many times per year. Although each COLEM organization may have several days or weeks between examination schedules, you may also apply to other COLEMS for additional testing dates. For example, one COLEM may have just scheduled a test in your area and the next date may be a month or more away.

However, you may request examination appointments from another COLEM! If you can afford to pay testing fees to several COLEMS, you could actually attempt the exam several times in a brief period of time, especially if you are willing to travel to another test site location which may be many miles away in another city or state. If your job or career depends upon the license - it might be worth the time, effort and expense.

CAUTION: You are NOT required to purchase additional training from these organizations. This is not necessary or required. Remember, all questions that will appear on any FCC examination are contained in this testing manual!

On the COLEM testing application, you should fill in the city or location where you want to be examined. You will be notified of a specific testing date, address, procedures, etc.

To make sure license applicants have paid all taxes and other obligations to the government, the FCC is required to obtain your taxpayer identification number, or Social Security number.

Anyone that owes the government money may be prevented from getting an FCC License because of the "US Debt Collection Act of 1996" (Big Brother is watching, making sure you have paid all government bills and past obligations). To keep your social security number confidential during the registration procedures, you can apply for and be issued a "FRN" - FCC Registration Number.

When applying for a new license, you have the option of obtaining an FRN by registering your Social Security or taxpayer ID number with the FCC in advance of being tested. Or, if you are not concerned - just give your Social Security number to the COLEM examiner (on FCC application Form 605).

You can obtain an FRN from the government website: www.fcc.gov. Click on "Commission Registration System," then give the FCC your Social Security number. The FCC will check your "government file" and if you do not own money to any federal agency, etc., they will give you an FRN number.

If you have a question or problem filling the FRN web page, contact the FCC directly at (202) 414-1250.

When you arrive at the testing location, the local exam proctor may verify your identity by inspecting your driver's license, passport, or other "photo" style identification.

The national organization will generate the actual examination from the same group of questions that appear in this guide. If you are tested for both Element 1 and 3 (recommended), a total of 124 questions will be randomly selected for your examination (24 for Element 1 and 100 for Element 3).

Your FCC exam should be graded locally and you should know if you passed at the test location. The test will be returned to the national COLEM headquarters. Although the examination process is now contracted to private testing administrators, the actual license is officially issued by the federal government FCC Licensing Division, in Gettysburg, Pennsylvania

How to Get Your Exam Appointment

If you are planning to take the FCC License exam in the near future, it is advisable to contract one of the following testing organizations to arrange for an examination appointment.

There are several private testing organizations that give FCC License examinations. However, we have selected only four to be included in this guidebook. After extensive research, we have determined the following organizations offer the best and most convenient testing services:

1. **The International Association of Radio, Telecommunications and Electromagnetics** (iNARTE), now within the organization "Exemplar Global". PO Box 602, Milwaukee, WI 53201-0602. 600 North Plankinton Avenue, Milwaukee, WI 53203-2914. Exam testing is available at more than 187 testing centers in all 50 states and at U.S. Military installations (DANTES) worldwide. Testing by appointment. $65 for up to two exam elements, $130 for all three. Call for schedule and exam registration information or email: fcc@exemplarglobal.org. Fast link to the FCC page on the organization website: http://www.inarte.org/fcc Contact: Christian Thornton. (888) 722-2440.

2. **Electronic Technicians Association International, Inc.**, 5 Depot Street, Greencastle, IN 46135-8024. All FCC Elements are tested at exam sites located in all states, including some overseas Military installations. Examinations are given by appointment. The cost is $50 for up to 3 FCC exam elements. Call for information or to schedule an exam appointment. Contact: Teresa Maher. (800) 288-3824 or (765) 653-8262. FAX: (765) 653-4301. Email: eta@eta-I.org Website: www.eta-I.org

3. **PSI Services, Inc.**, 16821 SE McGillivray, Suite 201, Vancouver, WA 98683. All FCC Elements are available by appointment from examiners located at over 350 locations in all 50 states, often in commercial airports. The cost is $60 for one element, $90 for two elements and $120 for all three if taken at one session. After passing the examination, you will be given a "Proof of Passing" document to send to the FCC for your license. This organization also tests for FAA Licenses, so they are able to arrange for FCC Radiotelephone testing on a regular basis with very fast appointment scheduling. Contact: Linda Small (800) 211-2753 or (360) 859-1277. FAX: (360) 891-0958. Email: Lsmall@psionline.com Website: www.psionline.com

4. **International Society of Certified Electronics Technicians** (ISCET), 3000 Landers Street, Suite A, Fort Worth, TX 76107-5642. FCC test Elements are given by appointment from examiners and at US Military bases in selected foreign countries. $50 for one element, $70 for two and $90 for all three. This organization also gives exams for their own ISCET certification., so be sure to specify the FCC License exam when making an appointment. Contact: Patricia Bohon. (800) 946-8201 or (817) 921-9101 ext. 116. FAX (817) 921-3741. Email: info@iscet.org Website: www.iscet.org

Study Tips

Test yourself frequently with the questions in this manual. Use your reference books to clear up learning difficulties in the questions missed.

For best results, cover up the letter for the correct answers shown on the far right side of each page. Searching through the wrong answers for the correct choice provides realistic practice for the FCC exam. Later, many students underline or high-lite the correct answers for final review sessions before the exam appointment.

When taking the actual FCC exam, it is suggested that you cover all the multiple-choice answers for each question. Then, try to recall the answer from memory before looking over the four choices. This method will help prevent you from being tricked by the three out of four answer choices.

Entire books have been written on the subject of how to take multiple-choice exams. Careful study of the exams in this manual should give you ample review of all possible deceptions to the correct answers. Remember, often the exact answer to a question is not one of the possible choices. So, select the nearest correct answer. Also, you may discover the correct answer to be: All of the above or None of the above, etc.

Here's a tip to remember - For those questions having an answer choice with "All of these or All of the above" you will discover that the "D" choice will be the correct answer 90% of the time! The opposite is true for "D" answers with a "None of these or None of the above" choice. "None" answers are wrong 90% of the time they appear! So, if you need to guess an answer on the FCC test, now you know the odds for a few of the questions.

Caution: Spelling errors may appear on FCC exams! The official exam questions in this manual are the exact word by word questions and answers obtained from the FCC. We have included all misspellings and typographical errors in this guide, because they may also appear on the FCC exam you receive. After studying the material in this manual, you should be ready for all questions on the FCC exam.

For best results, learn and understand how to solve math problems and electronic theory questions. Memorize: terminology, rules and regulations, diagrams, tolerances and definitions, etc. When solving math problems, keep your formulas and figuring in a clear logical order. All problems should be saved for future reference.

When reading the multiple-choice answers in this manual, keep in mind that the actual A, B, C and D answers may appear in a different order on the FCC exam you receive. The COLEM may change the order of answers, but not the actual wording. This manual contains newly revised updated test questions and answers for passing current FCC exams. You will need to know virtually all information in this guide to pass the FCC license exam.

Selective Study Program

You should know that each license exam includes many types of questions, ranging from easy to memorize FCC Rules and Regulations to more advanced electronic theory material.

The Key Topic pages were written by the FCC and were NOT arranged in order by complexity. So, learning the Key Topics in numerical order may not be the best study method. Often, the more difficult Key Topic material comes up before the more easy to learn questions.

On the actual FCC Exam you will be asked one question from each Key Topic, which always contains six possible questions (or about one Topic per page in this guide).

FCC Element 1: There are 24 Key Topics. You will be given one question from each Topic on your exam for a total of 24 questions. 18 answered correctly is the passing score.

FCC Element 3: There are 100 Key Topics. You will be given one question from each Topic on your exam for a total of 100 questions. 75 answered correctly is the passing score.

FCC Element 8: There are 50 Key Topics. You will be given one question from each Topic on your exam for a total of 50 questions. 38 answered correctly is the passing score.

At first glance it would appear that all questions can be learned with the same study effort. For most people, this is not true. Many questions can be easily understood and memorized, while other questions require more effort because of advanced content, math solution problems or electronic theory concepts, etc.

To help you to study the easiest test questions first in each Element, we have sorted and created a list of the material you should study in three "Study Question Groups". The first contains Introductory material, the next group contains Mid-Level questions and the third group contains more Advanced questions.

When you begin your study for an FCC Element exam, you should know that many students have benefitted with my suggestion of starting with Study Group #1. After this material is learned, proceed with group #2 which contains Mid-Level questions, then on to the Advanced questions. Remember, when these first two groups are learned, it is encouraging to understand that you will then know all the answers to enough questions to actually pass the FCC exam, because you have learned 75% of all FCC exam questions!

The final portion of the Key Topics to learn is Study Group #3, Advanced, which comprises 25% of the questions that will appear on your exam. You are permitted to miss 25% of the questions on an FCC exam and still get your license.

So, if your time is limited and you are having difficulty learning more advanced material, it is recommended that you concentrate on the more basic questions first. However, most of the questions in Element 1 are easier to learn than those of the other two Elements.

The Element 3 and 8 exam contains several questions that require mathematical solutions and a more advanced knowledge of electronics. Virtually all of these questions are placed into Study Group #3. The study groups were designed to help you to learn material without getting immersed into more advanced concepts too soon. But, to make sure you pass the actual exam, spend ample time learning as much as possible from Study Group #3 so you will have a cushion to fall back on should you miss a few questions from the first two study groups.

Or, study for each license by simply beginning with Key Topic #1 and learn FCC material in topic numerical order - from 1 to 24 for Element 1, 1 to 100 for Element 3 and from 1 to 50 for Element 8.

If you decide to study FCC questions and answers with the recommended method, Key Topics could be learned and reviewed in the following Study Groups order:

FCC Element 1 Key Topic Study Groups

Study Group #1 - Introductory - 10 Questions

1	6
2	7
3	8
4	9
5	10

Study Group #2 - Mid-Level - 8 Questions

11	18
12	19
13	20
17	21

Study Group #3 - Advanced - 6 Questions

14	22
15	23
16	24

FCC Element 3 Key Topic Study Groups

Study Group #1 - Introductory - 30 Questions

1	6	77	84	91	96
2	7	78	85	92	97
3	8	79	86	93	98
4	75	82	87	94	99
5	76	83	90	95	100

Study Group #2 - Mid-Level - 46 Questions

19	28	43	51	59	73
20	29	44	52	63	74
21	30	45	53	65	80
22	31	46	54	68	81
23	37	47	55	69	88
24	38	48	56	70	89
25	41	49	57	71	
26	42	50	58	72	

Study Group #3 - Advanced - 24 Questions

9	15	33	53
10	16	34	60
11	17	35	61
12	18	36	62
13	27	39	64
14	32	40	66

FCC Element 8 Key Topic Study Groups

Study Group #1 - Introductory - 19 Questions

11	24	31	47
12	27	33	48
13	28	41	49
19	29	42	50
21	30	43	

Study Group #2 - Mid-Level - 19 Questions

1	17	25	38
8	18	34	39
14	20	35	44
15	22	36	46
16	23	37	

Study Group #3 - Advanced - 12 Questions

2	9
3	10
4	26
5	32
6	40
7	45

Before Taking the FCC Exam

Get a full nights sleep. The benefits to be gained by a few extra hours of study are more than offset by the detrimental effects of fatigue.

Don't review or study on the day of the test. You might lose confidence in your ability and not do as well. Use of reference materials is not permitted. You should not bring any books, papers, notes or study guides to the examination.

Bring the following items with you to the testing site:

1. Photo identification; Driver's license, Passport, FRN Registration or Social Security Number, etc.

2. Pencils and ball point pens (blue or black ink).

3. Electronic Calculator. Only hand-held, battery operated models (not programmable) may be brought to the examination room. Be sure to ask the examiner how much time you will have to complete taking your test. In past years, there was a minimum of 3 hours time set aside per exam sitting.

4. Wear comfortable shoes and clothing. Most people spend several hours sitting on a hard uncomfortable chair while taking the examination. And remember, some examiners may not allow you to leave the testing room to go to the rest room during the exam.

Have a good mental attitude. You may miss 25% of the questions in each element and still pass the exam. Also, you should find the actual FCC examination no more difficult than the exams in this guidebook. On the following pages, you will find all the FCC multiple-choice questions that your examiner will select from to create your test.

Taking the FCC Exam

When you arrive for your appointment at the testing site, you will be provided all necessary FCC forms and test material. The testing fee charged by the COLEM examiner is usually collected in advance when you make your appointment or when you arrive at the testing location.

Next, you will fill out FCC Form 605 and Schedule E and specify which license(s) you are applying for. At this time, you can provide your Social Security number on the form. Or, if you have already applied for your FRN, you can simply provide your FRN number on FCC Form 605.

Take your time filling out all forms! Any errors on these forms may cause you to flunk the exam. If you have any questions on these forms, do not hesitate to ask the examiner for help.

Answer the exam questions in this order: (1) Answer easy questions first; rules and regulations, terms, etc. (2) Answer easiest theory questions next. (3) Solve easiest math problems. (4) Go back and solve the difficult questions you skipped.

When reading each FCC question, cover the answers with your answer sheet or other paper. Attempt to answer the question, THEN read each answer carefully. Remember, the multiple choice answers may be in a different order than in this manual.

Some of the math questions may NOT be exact figures, so you should select the nearest correct answer. All math problems are included on the following pages.

Answer EVERY question, even if you have to guess! There are only four multiple-choice answers for each test question. Remember, you have a 25% chance of guessing the correct answer.

After completion of the FCC exam, double check to see if you have marked every answer correctly. Since you will be skipping difficult questions in the first place, be sure you have answered all questions on the test and your have not skipped an answer choice.

Spelling errors on the FCC exam may occur. The exam questions in this manual contain the same misspellings, because we want you to be familiar with errors released by the federal government. Just because a question has a spelling error, or poor word usage, etc., it may still be correct. You will find all such errors reproduced on purpose in this guide so you will be comfortable selecting answers exactly as they appear on the federal exam. However, some examiners may make the effort to correct certain spelling errors - so be prepared to find either or both on your exam.

Please note: All schematic diagrams on the federal exam are included in this study guide. No attempt was made to alter or improve the original diagrams, so you will see exactly what will be on your exam.

After Passing the FCC Exam

If you successfully pass the FCC license examination, the COLEM test administrator should provide you a "Proof of Passing Certificate" (PPC). This form shows that you have passed and receive credit for the examination elements attempted. Be sure to keep your copy of the PPC or any other material at the test site that documents your successful passing of the exam.

The COLEM private testing service organization will NOT issue the official federal FCC license. The actual license is issued exclusively by the Federal Communications Commission. If for some reason the COLEM does not send in your materials to the FCC on your behalf, you will need to send a copy of your PPC Certificate along with FCC Form #605. This form actually contains two sections - Form #605 and another attachment. "Schedule E". Both of these forms must be filled out completely.

Some COLEMS may include this form with your PPC Certificate. If Form #605 is not provided, you may telephone the FCC at (202) 418-3676, or (800) 418-3676. FCC Forms are also available by downloading from the FCC's website at http://www.fcc.gov/formpage.html. Also, you may use the FCC Fax-On-Demand system by calling (202) 418-0177 from the handset of your fax machine (request the index to find the document number first).

Again, it is not usually necessary for you to send in your own completed application Form #605 and your PPC certificate to the FCC. Most COLEMs file these documents for applicants who test with them. Be sure to ask your COLEM if they provides this service. However, if you are told that it is your responsibility to send in the forms, use the following address to receive your license:

Federal Communications Commission
1270 Fairfield Road
Gettysburg, PA 17325-7245

If you need assistance filling out an FCC form, or have any questions, phone (202) 418-0680. It is highly recommended that you photocopy both the completed FCC Form #605 and your PPC certificate prior to mailing these documents to the FCC for license processing.

The FCC does not charge any fee for an FCC Radiotelephone Operator License or the RADAR Endorsement. You must pay a fee to take the examination, but the actual license is issued by the Federal Communications Commission.

FCC Licenses are issued for the "Lifetime". Good news for license applicants, the FCC General Radiotelephone Operator License is issued for life. No additional renewals or continuing education requirements are required, although additional training is often desirable and recommended.

If you have obtained an FCC License several years ago, lost your original license or had a legal name change, etc., you will need to apply for a changed, renewal or duplicate license. Use FCC Form 605 and the FCC Form 159 (fee processing form) to replace a lost, stolen or mutilated license.

Marine Radio & Restricted Operator Permits

MARINE RADIO OPERATOR PERMIT: Most readers of this guide will apply for the General Radiotelephone Operator License, which is FCC Elements 1 and 3 (There is no Element 2 - it was combined into Element 1 years ago). You must pass the Element 1 exam before your qualify to take the Element 3 exam (or have the Element 3 exam graded, if you are taking them both at the same time). However, if you are only interested in getting the FCC Marine Radio Permit you need only pass the Element 1 exam by correctly answering 18 out of 24 questions. (You many NOT hold both the Marine Permit and the General License).

Holders of this permit are FCC licensed to operate radiotelephone stations aboard certain vessels that sail the Great Lakes and elsewhere. They are also authorized to operate radio-telephone stations aboard vessels weighing more than 300 gross tons or which carry more than six passengers for hire in the open sea or any tidewater area of the United States. They are also licensed to operate certain aviation and coast radiotelephone stations. This FCC Permit is issued for the lifetime of the holder.

RESTRICTED RADIOTELEPHONE OPERATOR PERMIT: This permit does not require any examination and it has a very limited scope of authority. Holders of this permit are authorized to operate most aircraft and aeronautical ground radio stations. Also authorized is the operation of marine radiotelephone stations aboard craft carrying less than six passengers for hire in the open sea, on the Great Lakes or bays or tidewaters if operator licensing is required. It is most often used by aviation pilots, non-technical operators, etc. One person aboard stations in the maritime and aviation services must have this FCC Permit when making international flights, voyages or communications, when using frequencies below 30 MHz., when using satellite ship stations or when operating a vessel subject to the FCC Bridge to Bridge Act.

Because this FCC Permit requires a fee, you must obtain a FCC Registration Number (FRN) and supply it on forms when you are sending money to the Commission. You will need FCC Form #160. This can be obtained from the FCC Website listed on the previous page. After you have submitted Form #160 and received your FRN number you will then need to fill out and send in FCC Forms #605 (Main page and the Schedule E form) along with yet another document, FCC Form #159 which is the "Fee Remittance Advice" form. This FCC Permit has a one-time $50 fee. When filling out the fee form you will list "PACS" for the payment fee code type.

If you have questions regarding your application or fee, you may call the FCC's Consumer Center at (888) 225-5322.

OTHER FCC ELEMENTS AND LICENSES: Element 4 was phased out. Elements 5 and 6 are for Radiotelegraph stations that use Morse Code or for shipboard operators required to be telegraph licensed. Element 7 and 9 are for the CMDSS Operator and Maintenance licenses governing ships that sail in certain seaways. These licenses are primarily for repair and maintenance at sea for GMDSS equipment. Also, under certain circumstances the applicant may need to pass a U.S. Coast Guard approved course in addition to the GMDSS license. Due to the little demand for the GMDSS license it is not included in this guidebook. However, since GMDSS is covered by many of the same rules & regulations as the General Radiotelephone License, you will find many GMDSS related questions & answers in Element 1 and 3.

The main office of the FCC for general information is located at 445 12th Street SW, Washington, DC 20554. Information and inquiries can be emailed to fccinfo@fcc.gov. The primary phone number is (888) 225-5322 or (888)-CALL-FCC. Also, a Special Information phone number is (202) 418-0680. FAX: (202) 418-0232.

Radio Law & Operating Practice
FCC Element 1 Questions

Element 1 (formerly Element 1 & 2) is comprised of basic Radio Law and Operating Practice with which every commercial radio operator should be familiar. The exam contain 24 questions concerning provisions of laws, treaties, regulations, operating procedures and practices generally followed or required in communicating by means of radiotelephone stations. The minimum passing score is 18 questions answered correctly.

The 24 questions asked on your Element 1 examination will be taken from an FCC question pool of 144 questions covering four Subelements, each containing 36 possible questions. Your Element 1 exam will consist of one randomly selected multiple-choice question from each of the 24 "Key Topics" for a complete examination. You will be asked one question from each page in this section of the book. Almost every page has a complete Key Topic of six questions, so you will see one of these on your exam.

Although you may obtain a Marine Radio Operator Permit (MROP) after passing the Element 1 exam, most applicants prefer to continue on to Element 3 and obtain the FCC General Radiotelephone Operator License. You may NOT hold both a GROL and a MROP at the same time. Since the GROL is a higher class license, most applicants test for Element 3 immediately after taking the Element 1 exam.

Since the Element 1 exam is easier than the FCC Element 3 exam, most applicants are successful on the first attempt to pass this test. If you sit for both the Element 1 and Element 3 exam and pass only Element 1, you quality for the FCC Marine Radio Operator Permit (MROP).

Some people who pass the Element 1 exam have benefitted by getting the FCC Marine Radio Operator Permit Certificate to show to an employer or school to prove success and progress in passing the prerequisite stage to the FCC General Radiotelephone License. For others, the Marine Radio Operator Permit may be the only license necessary for job, civic, military or education requirements.

Contents of the Element 1 Examination:

FCC Commercial Element 1 Questions

Subelement A - Rules & Regulations: 6 Key Topics - 6 Exam Questions

Key Topic 1: Equipment Requirements

1. What is a requirement of all marine transmitting apparatus used aboard United States vessels? **A**
   - A. Only equipment that has been certified by the FCC for Part 80 operations is authorized.
   - B. Equipment must be type-accepted by the U.S. Coast Guard for maritime mobile use.
   - C. Certification is required by the International Maritime Organization (IMO).
   - D. Programming of all maritime channels must be performed by a licensed Marine Radio Operator.

2. What transmitting equipment is authorized for use by a station in the maritime services? **B**
   - A. Transmitters that have been certified by the manufacturer for maritime use.
   - B. Unless specifically excepted, only transmitters certified by the Federal Communications Commission for Part 80 operations.
   - C. Equipment that has been inspected and approved by the U.S. Coast Guard.
   - D. Transceivers and transmitters that meet all ITU specifications for use in maritime mobile service.

3. Small passenger vessels that sail 20 to 150 nautical miles from the nearest land must have what additional equipment? **D**
   - A. Inmarsat-B terminal.
   - B. Inmarsat-C terminal.
   - C. Aircraft Transceiver with 121.5 MHz.
   - D. MF-HF SSB Transceiver.

4. What equipment is programmed to initiate transmission of distress alerts and calls to individual stations? **C**
   - A. NAVTEX.
   - B. GPS.
   - C. DSC controller.
   - D. Scanning Watch Receiver.

5. What is the minimum transmitter power level required by the FCC for a medium-frequency transmitter aboard a compulsorily fitted vessel? **B**
   - A. At least 100 watts, single-sideband, suppressed-carrier power.
   - B. At least 60 watts PEP.
   - C. The power predictably needed to communicate with the nearest public coast station operating on 2182 kHz.
   - D. At least 25 watts delivered into 50 ohms effective resistance when operated with a primary voltage of 13.6 volts DC.

6. Shipboard transmitters using F3E emission (FM voice) may not exceed what carrier power? **D**
   - A. 500 watts.
   - B. 250 watts.
   - C. 100 watts.
   - D. 25 watts.

Key Topic 2: License Requirements

1. Which commercial radio operator license is required to operate a fixed-tuned ship RADAR station with external controls? **D**
   - A. A radio operator certificate containing a Ship RADAR Endorsement.
   - B. A Marine Radio Operator Permit or higher.
   - C. Either a First or Second Class Radiotelegraph certificate or a General Radiotelephone Operator License.
   - D No radio operator authorization is required.

2. When is a Marine Radio Operator Permit or higher license required for aircraft communications? **C**
   - A. When operating on frequencies below 30 MHz allocated exclusively to aeronautical mobile services.
   - B. When operating on frequencies above 30 MHz allocated exclusively to aeronautical mobile services.
   - C. When operating on frequencies below 30 MHz not allocated exclusively to aeronautical mobile services.
   - D. When operating on frequencies above 30 MHz not assigned for international use.

3. Which of the following persons are ineligible to be issued a commercial radio operator license? **A**
   - A. Individuals who are unable to send and receive correctly by telephone spoken messages in English.
   - B. Handicapped persons with uncorrected disabilities which affect their ability to perform all duties required of commercial radio operators.
   - C. Foreign maritime radio operators unless they are certified by the International Maritime Organization (IMO).
   - D. U.S. Military radio operators who are still on active duty.

4. What are the radio operator requirements of a passenger ship equipped with a GMDSS installation? **D**
   - A. The operator must hold a General Radiotelephone Operator License or higher-class license.
   - B. The operator must hold a Restricted Radiotelephone Operator Permit or higher-class license.
   - C. The operator must hold a Marine Radio Operator Permit or higher-class license.
   - D. Two operators on board must hold a GMDSS Radio Operator License or a Restricted GMDSS Radio Operator License, depending on the ship's operating areas.

5 What is the minimum radio operator requirement for ships subject to the Great Lakes Radio Agreement? **C**
   - A. Third Class Radiotelegraph Operator's Certificate.
   - B. General Radiotelephone Operator License.
   - C. Marine Radio Operator Permit.
   - D. Restricted Radiotelephone Operator Permit.

6. What is a requirement of every commercial operator on duty and in charge of a transmitting system? **B**
   - A. A copy of the Proof-of-Passing Certificate (PPC) must be in the station's records.
   - B. The original license or a photocopy must be posted or in the operator's personal possession and available for inspection.
   - C. The FCC Form 605 certifying the operator's qualifications must be readily available at the transmitting system site.
   - D. A copy of the operator's license must be supplied to the radio station's supervisor as evidence of technical qualification.

Key Topic 3: Watchkeeping

1. Radio watches for compulsory radiotelephone stations will include the following: **C**
   A. VHF channel 22a continuous watch at sea.
   B. 121.5 MHz continuous watch at sea.
   C. VHF channel 16 continuous watch.
   D. 500 kHz.

2. All compulsory equipped cargo ships (except those operating under GMDSS regulations or in a VTS) while being navigated outside of a harbor or port, shall keep a continuous radiotelephone watch on: **A**
   A. 2182 kHz and Ch-16.
   B. 2182 kHz.
   C. Ch-16.
   D. Cargo ships are exempt from radio watch regulations.

3. What channel must all compulsory, non-GMDSS vessels monitor at all times in the open sea? **D**
   A. Channel 8.
   B. Channel 70.
   C. Channel 6.
   D. Channel 16.

4. When a watch is required on 2182 kHz, at how many minutes past the hour must a 3 minute silent period be observed? **A**
   A. 00, 30.
   B. 15, 45.
   C. 10, 40.
   D. 05, 35.

5. Which is true concerning a required watch on VHF Ch-16? **D**
   A. It is compulsory at all times while at sea until further notice, unless the vessel is in a VTS system.
   B. When a vessel is in an A1 sea area and subject to the Bridge-to-Bridge act and in a VTS system, a watch is not required on Ch-16, provided the vessel monitors both Ch-13 and VTS channel.
   C. It is always compulsory in sea areas A2, A3 and A4.
   D. All of the above.

6. What are the mandatory DSC watchkeeping bands/channels? **B**
   A. VHF Ch-70, 2 MHz MF DSC, 6 MHz DSC and 1 other HF DSC.
   B. 8 MHz HF DSC, 1 other HF DSC, 2 MHz MF DSC and VHF Ch-70.
   C. 2 MHz MF DSC, 8 MHz DSC, VHF Ch-16 and 1 other HF DSC.
   D. None of the above.

Key Topic 4: Logkeeping

1. Who is required to make entries in a required service or maintenance log? **B**
   - A. The licensed operator or a person whom he or she designates.
   - B. The operator responsible for the station operation or maintenance.
   - C. Any commercial radio operator holding at least a Restricted Radiotelephone Operator Permit.
   - D. The technician who actually makes the adjustments to the equipment.

2. Who is responsible for the proper maintenance of station logs? **D**
   - A. The station licensee.
   - B. The commercially-licensed radio operator in charge of the station.
   - C. The ship's master and the station licensee.
   - D. The station licensee and the radio operator in charge of the station.

3. Where must ship station logs be kept during a voyage? **A**
   - A. At the principal radiotelephone operating position.
   - B. They must be secured in the vessel's strongbox for safekeeping.
   - C. In the personal custody of the licensed commercial radio operator.
   - D. All logs are turned over to the ship's master when the radio operator goes off duty.

4. What is the proper procedure for making a correction in the station log? **C**
   - A. The ship's master must be notified, approve and initial all changes to the station log.
   - B. The mistake may be erased and the correction made and initialized only by the radio operator making the original error.
   - C. The original person making the entry must strike out the error, initial the correction and indicate the date of the correction.
   - D. Rewrite the new entry in its entirety directly below the incorrect notation and initial the change.

5. How long should station logs be retained when there are entries relating to distress or disaster situations? **B**
   - A. Until authorized by the Commission in writing to destroy them.
   - B. For a period of three years from the last date of entry, unless notified by the FCC.
   - C. Indefinitely, or until destruction is specifically authorized by the U.S. Coast Guard.
   - D. For a period of one year from the last date of entry.

6. How long should station logs be retained when there are no entries relating to distress or disaster situations? **C**
   - A. For a period of three years from the last date of entry, unless notified by the FCC.
   - B. Until authorized by the Commission in writing to destroy them.
   - C. For a period of two years from the last date of entry.
   - D. Indefinitely, or until destruction is specifically authorized by the U.S. Coast Guard.

Key Topic 5: Log Entries

1. Radiotelephone stations required to keep logs of their transmissions must include: **D**
   - A. Station, date and time.
   - B. Name of operator on duty.
   - C. Station call signs with which communication took place.
   - D. All of these.

2. Which of the following is true? **B**
   - A. Battery test must be logged daily.
   - B. EPIRB tests are normally logged monthly.
   - C. Radiotelephone tests are normally logged weekly.
   - D. None of the above.

3. Where should the GMDSS radio log be kept on board ship? **C**
   - A. Captain's office.
   - B. Sea cabin.
   - C. At the GMDSS operating position.
   - D. Anywhere on board the vessel.

4. Which of the following statements is true? **A**
   - A. Key letters or abbreviations may be used in GMDSS Radio Logbooks if their meaning is noted in the log.
   - B. Key letters or abbreviations may not be used in GMDSS Radio Logbooks under any circumstances.
   - C. All Urgency communications must be entered in the logbook.
   - D. None of the above.

5. Which of the following logkeeping statements is true? **B**
   - A. Entries relating to pre-voyage, pre-departure and daily tests are required.
   - B. Both a) and c)
   - C. A summary of all required Distress communications heard and Urgency communications affecting the station's own ship. Also, all Safety communications (other than VHF) affecting the station's own ship must be logged.
   - D. Routine daily MF-HF and Inmarsat-C transmissions do not have to be logged.

6. Which of the following statements concerning log entries is false? **A**
   - A. All Safety communications received on VHF must be logged.
   - B. All required equipment tests must be logged.
   - C. The radio operator must log on and off watch.
   - D. The vessels daily position must be entered in the log.

Key Topic 6: Miscellaneous Rules & Regulations

1. What regulations govern the use and operation of FCC-licensed ship stations in international waters? **B**
   - A. The regulations of the International Maritime Organization (IMO) and Radio Officers Union.
   - B. Part 80 of the FCC Rules plus the international Radio Regulations and agreements to which the United States is a party.
   - C. The Maritime Mobile Directives of the International Telecommunication Union.
   - D. Those of the FCC's Wireless Telecommunications Bureau, Maritime Mobile Service, Washington, DC 20554.

2. When may the operator of a ship radio station allow an unlicensed person to speak over the transmitter? **C**
   - A. At no time. Only commercially-licensed radio operators may modulate the transmitting apparatus.
   - B. When the station power does not exceed 200 watts peak envelope power.
   - C. When under the supervision of the licensed operator.
   - D. During the hours that the radio officer is normally off duty.

3. Where do you make an application for inspection of a ship GMDSS radio station? **C**
   - A. To a Commercial Operator Licensing Examination Manager (COLE Manager).
   - B. To the Federal Communications Commission, Washington, DC 20554.
   - C. To the Engineer-in-Charge of the FCC District Office nearest the proposed place of inspection.
   - D. To an FCC-licensed technician holding a GMDSS Radio Maintainer's License.

4. Who has ultimate control of service at a ship's radio station? **A**
   - A. The master of the ship.
   - B. A holder of a First Class Radiotelegraph Certificate with a six months' service endorsement.
   - C. The Radio Officer-in-Charge authorized by the captain of the vessel.
   - D. An appointed licensed radio operator who agrees to comply with all Radio Regulations in force.

5. Where must the principal radiotelephone operating position be installed in a ship station? **C**
   - A. At the principal radio operating position of the vessel.
   - B. In the chart room, master's quarters or wheel house.
   - C. In the room or an adjoining room from which the ship is normally steered while at sea.
   - D. At the level of the main wheel house or at least one deck above the ship's main deck.

6. By international agreement, which ships must carry radio equipment for the safety of life at sea? **D**
   - A. All ships traveling more than 100 miles out to sea.
   - B. Cargo ships of more than 100 gross tons and passenger vessels on international deep-sea voyages.
   - C. All cargo ships of more than 100 gross tons.
   - D. Cargo ships of more than 300 gross tons and vessels carrying more than 12 passengers.

Subelement B - Communications Procedures: 6 Key Topics - 6 Exam Questions

Key Topic 7: Bridge-to-Bridge Operations

1. What traffic management service is operated by the U.S. Coast Guard in certain designated water areas to prevent ship collisions, groundings and environmental harm? **B**
   A. Water Safety Management Bureau (WSMB).
   B. Vessel Traffic Service (VTS).
   C. Ship Movement and Safety Agency (SMSA).
   D. Interdepartmental Harbor and Port Patrol (IHPP).

2. What is a bridge-to-bridge station? **D**
   A. An internal communications system linking the wheel house with the ship's primary radio operating position and other integral ship control points.
   B. An inland waterways and coastal radio station serving ship stations operating within the United States.
   C. A portable ship station necessary to eliminate frequent application to operate aship station on board different vessels.
   D. A VHF radio station located on a ship's navigational bridge or main control station that is used only for navigational communications.

3. When may a bridge-to-bridge transmission be more than 1 watt? **A**
   A. When broadcasting a distress message and rounding a bend in a river or traveling in a blind spot.
   B. When broadcasting a distress message.
   C. When rounding a bend in a river or traveling in a blind spot.
   D. When calling the Coast Guard.

4. When is it legal to transmit high power on Channel 13? **D**
   A. Failure of vessel being called to respond.
   B. In a blind situation such as rounding a bend in a river.
   C. During an emergency.
   D. All of these.

5. A ship station using VHF bridge-to-bridge Channel 13: **A**
   A. May be identified by the name of the ship in lieu of call sign.
   B. May be identified by call sign and country of origin.
   C. Must be identified by call sign and name of vessel.
   D. Does not need to identify itself within 100 miles from shore.

6. The primary purpose of bridge-to-bridge communications is: **C**
   A. Search and rescue emergency calls only.
   B. All short-range transmission aboard ship.
   C. Navigational communications.
   D. Transmission of Captain's orders from the bridge.

Key Topic 8: Operating Procedures-1

1. What is the best way for a radio operator to minimize or prevent interference to other stations? **C**
   - A. By using an omni-directional antenna pointed away from other stations.
   - B. Reducing power to a level that will not affect other on-frequency communications.
   - C. Determine that a frequency is not in use by monitoring the frequency before transmitting.
   - D. By changing frequency when notified that a radiocommunication causes interference.

2. Under what circumstances may a coast station using telephony transmit a general call to a group of vessels? **B**
   - A. Under no circumstances.
   - B. When announcing or preceding the transmission of Distress, Urgency, Safety or other important messages.
   - C. When the vessels are located in international waters beyond 12 miles.
   - D. When identical traffic is destined for multiple mobile stations within range.

3. Who determines when a ship station may transmit routine traffic destined for a coast or government station in the maritime mobile service? **C**
   - A. Shipboard radio officers may transmit traffic when it will not interfere with ongoing radiocommunications.
   - B. The order and time of transmission and permissible type of message traffic is decided by the licensed on-duty operator.
   - C. Ship stations must comply with instructions given by the coast or government station.
   - D. The precedence of conventional radiocommunications is determined by FCC and international regulation.

4. What is required of a ship station which has established initial contact with another station on 2182 kHz or Ch-16? **A**
   - A. The stations must change to an authorized working frequency for the transmission of messages.
   - B. The stations must check the radio channel for Distress, Urgency and Safety calls at least once every ten minutes.
   - C. Radiated power must be minimized so as not to interfere with other stations needing to use the channel.
   - D. To expedite safety communications, the vessels must observe radio silence for two out of every fifteen minutes.

5. How does a coast station notify a ship that it has a message for the ship? **D**
   - A. By making a directed transmission on 2182 kHz or 156.800 MHz.
   - B. The coast station changes to the vessel's known working frequency.
   - C. By establishing communications using the eight-digit maritime mobile service identification.
   - D. The coast station may transmit, at intervals, lists of call signs in alphabetical order for which they have traffic.

6. What is the priority of communications? **B**
   - A. Safety, Distress, Urgency and radio direction-finding.
   - B. Distress, Urgency and Safety.
   - C. Distress, Safety, radio direction-finding, search and rescue.
   - D. Radio direction-finding, Distress and Safety.

Key Topic 9: Operating Procedures-2

1. Under what circumstances may a ship or aircraft station interfere with a public coast station? **A**

   A. In cases of distress.
   B. Under no circumstances during on-going radiocommunications.
   C. During periods of government priority traffic handling.
   D. When it is necessary to transmit a message concerning the safety of navigation or important meteorological warnings.

2. Ordinarily, how often would a station using a telephony emission identify? **B**

   A. At least every 10 minutes.
   B. At the beginning and end of each transmission and at 15-minute intervals.
   C. At 15-minute intervals, unless public correspondence is in progress.
   D. At 20-minute intervals.

3. When using a SSB station on 2182 kHz or VHF-FM on channel 16: **D**

   A. Preliminary call must not exceed 30 seconds.
   B. If contact is not made, you must wait at least 2 minutes before repeating the call.
   C. Once contact is established, you must switch to a working frequency.
   D. All of these.

4. What should a station operator do before making a transmission? **A**

   A. Except for the transmission of distress calls, determine that the frequency is not in use by monitoring the frequency before transmitting.
   B. Transmit a general notification that the operator wishes to utilize the channel.
   C. Check transmitting equipment to be certain it is properly calibrated.
   D. Ask if the frequency is in use.

5. On what frequency should a ship station normally call a coast station when using a radiotelephony emission? **B**

   A. On a vacant radio channel determined by the licensed radio officer.
   B. Calls should be initiated on the appropriate ship-to-shore working frequency of the coast station.
   C. On any calling frequency internationally approved for use within ITU Region 2.
   D. On 2182 kHz or Ch-16 at any time.

6. In the International Phonetic Alphabet, the letters E, M, and S are represented by the words: **C**

   A. Echo, Michigan, Sonar.
   B. Equator, Mike, Sonar.
   C. Echo, Mike, Sierra
   D. Element, Mister, Scooter

Key Topic 10: Distress Communications

1. What information must be included in a Distress message? **D**

   A. Name of vessel.
   B. Location.
   C. Type of distress and specifics of help requested.
   D. All of the above.

2. What are the highest priority communications from ships at sea? **C**

   A. All critical message traffic authorized by the ship's master.
   B. Navigation and meteorological warnings.
   C. Distress calls are highest and then communications preceded by Urgency and then Safety signals.
   D. Authorized government communications for which priority right has been claimed.

3. What is a Distress communication? **B**

   A. Communications indicating that the calling station has a very urgent message concerning safety.
   B. An internationally recognized communication indicating that the sender is threatened by grave and imminent danger and requests immediate assistance.
   C. Radio communications which, if delayed, will adversely affect the safety of life or property.
   D. An official radio communication notification of approaching navigational or meteorological hazards.

4. What is the order of priority of radiotelephone communications in the maritime services? **C**

   A. Alarm and health and welfare communications.
   B. Navigation hazards, meteorological warnings, priority traffic.
   C. Distress calls and signals, followed by communications preceded by Urgency and Safety signals and all other communications.
   D. Government precedence, messages concerning safety of life and protection of property, and traffic concerning grave and imminent danger.

5. The radiotelephone Distress call and message consists of: **D**

   A. MAYDAY spoken three times, followed by the name of the vessel and the call sign in phonetics spoken three times.
   B. Particulars of its position, latitude and longitude, and other information which might facilitate rescue, such as length, color and type of vessel, and number of persons on board.
   C. Nature of distress and kind of assistance required.
   D. All of the above.

6. What is Distress traffic? **A**

   A. All messages relative to the immediate assistance required by a ship, aircraft or other vehicle threatened by grave or imminent danger, such as life and safety of persons on board, or man overboard.
   B. In radiotelephony, the speaking of the word, "Mayday."
   C. Health and welfare messages concerning property and the safety of a vessel.
   D. Internationally recognized communications relating to important situations.

Key Topic 11: Urgency and Safety Communications

1. What is a typical Urgency transmission? **A**
   - A. A request for medical assistance that does not rise to the level of a Distress or a critical weather transmission higher than Safety.
   - B. A radio Distress transmission affecting the security of humans or property.
   - C. Health and welfare traffic which impacts the protection of on-board personnel.
   - D. A communications alert that important personal messages must be transmitted.

2. What is the internationally recognized Urgency signal? **B**
   - A. The letters "TTT" transmitted three times by radiotelegraphy.
   - B. The words "PAN PAN" spoken three times before the Urgency call.
   - C. Three oral repetitions of the word "Safety" sent before the call.
   - D. The pronouncement of the word "Mayday."

3. What is a Safety transmission? **A**
   - A. A communications transmission which indicates that a station is preparing to transmit an important navigation or weather warning.
   - B. A radiotelephony warning preceded by the words "PAN PAN."
   - C. Health and welfare traffic concerning the protection of human life.
   - D. A voice call proceeded by the words "Safety Alert."

4. The Urgency signal concerning the safety of a ship, aircraft or person shall be sent only on the authority of: **C**
   - A. Master of ship.
   - B. Person responsible for mobile station.
   - C. Either Master of ship or person responsible for mobile station.
   - D. An FCC-licensed operator.

5. The Urgency signal has lower priority than: **B**
   - A. Ship-to-ship routine calls.
   - B. Distress.
   - C. Safety.
   - D. Security.

6. What safety signal call word is spoken three times, followed by the station call letters spoken three times, to announce a storm warning, danger to navigation, or special aid to navigation? **D**
   - A. PAN PAN.
   - B. MAYDAY.
   - C. SAFETY.
   - D. SECURITY*.

*May be inaccurately spelled "SECURITE" on the FCC examination

Key Topic 12: GMDSS

1. What is the fundamental concept of the GMDSS? **D**
   - A. It is intended to automate and improve existing digital selective calling procedures and techniques.
   - B. It is intended to provide more effective but lower cost commercial communications.
   - C. It is intended to provide compulsory vessels with a collision avoidance system when they are operating in waters that are also occupied by non-compulsory vessels.
   - D. It is intended to automate and improve emergency communications in the maritime industry.

2. The primary purpose of the GMDSS is to: **C**
   - A. Allow more effective control of SAR situations by vessels.
   - B. Provide additional shipboard systems for more effective company communications.
   - C. Automate and improve emergency communications for the world's shipping industry.
   - D. Provide effective and inexpensive communications.

3. What is the basic concept of GMDSS? **D**
   - A. Shoreside authorities and vessels can assist in a coordinated SAR operation with minimum delay.
   - B. Search and rescue authorities ashore can be alerted to a Distress situation.
   - C. Shipping in the immediate vicinity of a ship in Distress will be rapidly alerted.
   - D. All of these.

4. GMDSS is primarily a system based on? **D**
   - A. Ship-to-ship Distress communications using MF or HF radiotelephony.
   - B. VHF digital selective calling from ship to shore.
   - C. Distress, Urgency and Safety communications carried out by the use of narrow-band direct printing telegraphy.
   - D. The linking of search and rescue authorities ashore with shipping in the immediate vicinity of a ship in Distress or in need of assistance.

5. What is the responsibility of vessels under GMDSS? **C**
   - A. Vessels over 300 gross tons may be required to render assistance if such assistance does not adversely affect their port schedule.
   - B. Only that vessel, regardless of size, closest to a vessel in Distress, is required to render assistance.
   - C. Every ship is able to perform those communications functions that are essential for the Safety of the ship itself and of other ships.
   - D. Vessels operating under GMDSS, outside of areas effectively serviced by shoreside authorities, operating in sea areas A2, and A4 may be required to render assistance in Distress situations.

6. GMDSS is required for which of the following? **B**
   - A. All vessels capable of international voyages.
   - B. SOLAS Convention ships of 300 gross tonnage or more.
   - C. Vessels operating outside of the range of VHF coastal radio stations.
   - D. Coastal vessels of less than 300 gross tons.

Subelement C - Equipment Operations: 6 Key Topics - 6 Exam Questions

Key Topic 13: VHF Equipment Controls

1. What is the purpose of the INT-USA control settings on a VHF? **C**

   A. To change all VTS frequencies to Duplex so all vessels can receive maneuvering orders.
   B. To change all VHF channels from Duplex to Simplex while in U.S. waters.
   C. To change certain International Duplex channel assignments to simplex in the U.S. for VTS and other purposes.
   D. To change to NOAA weather channels and receive weather broadcasts while in the U.S.

2. VHF ship station transmitters must have the capability of reducing carrier power to: **A**

   A. 1 watt.
   B. 10 watts.
   C. 25 watts.
   D. 50 watts.

3. The Dual Watch (DW) function is used to: **D**

   A. Listen to Ch-70 at the same time while monitoring Ch-16.
   B. Sequentially monitor 4 different channels.
   C. Sequentially monitoring all VHF channels.
   D. Listen on any selected channel while periodically monitoring Ch-16.

4. Which of the following statements best describes the correct setting for manual adjustment of the squelch control? **A**

   A. Adjust squelch control to the minimum level necessary to barely suppress any background noise.
   B. Always adjust squelch control to its maximum level.
   C. Always adjust squelch control to its minimum level.
   D. Adjust squelch control to approximately twice the minimum level necessary to barely suppress any background noise.

5. The "Scan" function is used to: **D**

   A. Monitor Ch-16 continuously and switching to either Ch-70 or Ch-13 every 5 seconds.
   B. Scan Ch-16 for Distress calls.
   C. Scan Ch-70 for Distress alerts.
   D. Sequentially scan all or selected channels.

6. Why must all VHF Distress, Urgency and Safety communications (as well as VTS traffic calls) be performed in Simplex operating mode? **B**

   A. To minimize interference from vessels engaged in routine communications.
   B. To ensure that vessels not directly participating in the communications can hear both sides of the radio exchange.
   C. To enable an RCC or Coast station to only hear communications from the vessel actually in distress.
   D. To allow an RCC or Coast station to determine which transmissions are from other vessels and which transmissions are from the vessel actually in distress.

Key Topic 14: VHF Channel Selection

1. What channel must VHF-FM-equipped vessels monitor at all times when the vessel is at sea? **B**
   - A. Channel 8.
   - B. Channel 16.
   - C. Channel 5A.
   - D. Channel 1A.

2. What is the aircraft frequency and emission used for distress communications? **D**
   - A. 243.000 MHz - F3E.
   - B. 121.500 MHz - F3E.
   - C. 156.525 MHz - F1B.
   - D. 121.500 MHz - A3E.

3. Which VHF channel is used only for digital selective calling? **A**
   - A. Channel 70.
   - B. Channel 16.
   - C. Channel 22A.
   - D. Channel 6.

4. Which channel is utilized for the required bridge-to-bridge watch? **C**
   - A. DSC on Ch-70.
   - B. VHF-FM on Ch-16.
   - C. VHF-FM on Ch-13 in most areas of the continental United States.
   - D. The vessel's VHF working frequency.

5. Which channel would most likely be used for routine ship-to-ship voice traffic? **B**
   - A. Ch-16.
   - B. Ch-08.
   - C. Ch-70.
   - D. Ch-22A.

6. What channel would you use to place a call to a shore telephone? **C**
   - A. Ch-16.
   - B. Ch-70.
   - C. Ch-28.
   - D. Ch-06.

Key Topic 15: MF-HF Equipment Controls

1. Which modes could be selected to receive vessel traffic lists from high seas shore stations? **D**

   A. AM and VHF-FM.
   B. ARQ and FEC.
   C. VHF-FM and SSB.
   D. SSB and FEC.

2. Why must all MF-HF Distress, Urgency and Safety communications take place solely on the 6 assigned frequencies and in the simplex operating mode? **B**

   A. For non-GMDSS ships, to maximize the chances for other vessels to receive those communications.
   B. Answers a) and c) are both correct.
   C. For GMDSS or DSC-equipped ships, to maximize the chances for other vessels to receive those communications following the transmission of a DSC call of the correct priority.
   D. To enable an RCC or Coast station to only hear communications from the vessel actually in distress.

3. To set-up the MF/HF transceiver for a voice call to a coast station, the operator must: **C**

   A. Select J3E mode for proper SITOR operations.
   B. Select F1B mode or J2B mode, depending on the equipment manufacturer.
   C. Select J3E mode for proper voice operations.
   D. Select F1B/J2B modes or J3E mode, depending on whether FEC or ARQ is preferred.

4. MF/HF transceiver power levels should be set: **A**

   A. To the lowest level necessary for effective communications.
   B. To the level necessary to maximize the propagation radius.
   C. To the highest level possible so as to ensure other stations cannot "break-in"on the channel during use.
   D. Both a) and c) are correct.

5. To set-up the MF/HF transceiver for a TELEX call to a coast station, the operator must: **B**

   A. Select J3E mode for proper SITOR operations.
   B. Select F1B mode or J2B mode, depending on the equipment manufacturer.
   C. Select F1B/J2B modes or J3E mode, depending on whether ARQ or FEC is preferred.
   D. None of the above.

6. What is the purpose of the Receiver Incremental Tuning (RIT) or "Clarifier" control? **A**

   A. It acts as a "fine-tune" control on the receive frequency.
   B. It acts as a "fine-tune" control on the transmitted frequency.
   C. It acts as a "fine-tune" control on both the receive and transmitted frequencies.
   D. None of the above.

Key Topic 16: MF-HF Frequency & Emission Selection

1. On what frequency would a vessel normally call another ship station when using a radiotelephony emission? **B**
   - A. Only on 2182 kHz in ITU Region 2.
   - B. On 2182 kHz or Ch-16, unless the station knows that the called vessel maintains a simultaneous watch on another intership working frequency.
   - C. On the appropriate calling channel of the ship station at 15 minutes past the hour.
   - D. On the vessel's unique working radio channel assigned by the Federal Communications Commission.

2. What is the MF radiotelephony calling and Distress frequency? **C**
   - A. 2670 kHz.
   - B. Ch-06 VHF.
   - C. 2182 kHz.
   - D. Ch-22 VHF.

3. For general communications purposes, paired frequencies are: **D**
   - A. Normally used with private coast stations.
   - B. Normally used between ship stations.
   - C. Normally used between private coast and ship stations.
   - D. Normally used with public coast stations.

4. What emission must be used when operating on the MF distress and calling voice frequency? **A**
   - A. J3E - Single sideband telephony.
   - B. A1A - On-off keying without modulation by an audio frequency.
   - C. F3E - Frequency modulation telephony.
   - D. A3E - Amplitude modulation telephony, double sideband.

5. Which of the following defines high frequency "ITU Channel 1212"? **C**
   - A. Ch-12 in the 16 MHz band.
   - B. Ch-1216 in the MF band.
   - C. The 12th channel in the 12 MHz band.
   - D. This would indicate the 1st channel in the 12 MHz band.

6. For general communications purposes, simplex frequencies are: **D**
   - A. Normally used between ship stations and private coast stations.
   - B. Normally used with public coast stations.
   - C. Normally used between ship stations.
   - D. Both a) and c) are correct.

Key Topic 17: Equipment Tests

1. What is the proper procedure for testing a radiotelephone installation? **B**
   - A. A dummy antenna must be used to insure the test will not interfere with ongoing communications.
   - B. Transmit the station's call sign, followed by the word "test" on the frequency being used for the test.
   - C. Permission for the voice test must be requested and received from the nearest public coast station.
   - D. Short tests must be confined to a single frequency and must never be conducted in port.

2. When testing is conducted on 2182 kHz or Ch-16, testing should not continue for more than ______ in any 5-minute period. **D**
   - A. 2 minutes.
   - B. 1 minute.
   - C. 30 seconds.
   - D. 10 seconds.

3. Under GMDSS, a compulsory VHF-DSC radiotelephone installation must be tested at what minimum intervals at sea? **A**
   - A. Daily.
   - B. Annually, by a representative of the FCC.
   - C. Weekly.
   - D. Monthly.

4. The best way to test the MF-HF NBDP system is? **D**
   - A. Make a radiotelephone call to a coast station.
   - B. Initiate an ARQ call to demonstrate that the transmitter and antenna are working.
   - C. Initiate an FEC call to demonstrate that the transmitter and antenna are working.
   - D. Initiate an ARQ call to a Coast Station and wait for the automatic exchange of answerbacks.

5. The best way to test the Inmarsat-C terminal is? **A**
   - A. Compose and send a brief message to your own Inmarsat-C terminal.
   - B. Send a message to a shore terminal and wait for confirmation.
   - C. Send a message to another ship terminal.
   - D. If the "Send" light flashes, proper operation has been confirmed.

6. When may you test a radiotelephone transmitter on the air? **C**
   - A. Between midnight and 6:00 AM local time.
   - B. Only when authorized by the Commission.
   - C. At any time (except during silent periods) as necessary to assure proper operation.
   - D. After reducing transmitter power to 1 watt.

Key Topic 18: Equipment Faults

1. Under normal circumstances, what do you do if the transmitter aboard your ship is operating off-frequency, overmodulating or distorting? **C**
   - A. Reduce to low power.
   - B. Reduce audio volume level.
   - C. Stop transmitting.
   - D. Make a notation in station operating log.

2. Which would be an indication of proper operation of a SSB transmitter rated at 60 watt PEP output? **B**
   - A. In SSB (J3E) voice mode, with the transmitter keyed but without speaking into the microphone, power output is indicated.
   - B. In SITOR communications, the power meter can be seen fluctuating regularly from zero to the 60 watt relative output reading.
   - C. In SSB (J3E) mode, speaking into the microphone causes power meter to fluctuate slightly around the 60 watt reading.
   - D. A steady indication of transmitted energy on an RF Power meter with no fluctuations when speaking into the microphone.

3. If a ship radio transmitter signal becomes distorted: **C**
   - A. Reduce transmitter power.
   - B. Use minimum modulation.
   - C. Cease operations.
   - D. Reduce audio amplitude.

4. What would be an indication of a malfunction on a GMDSS station with a 24 VDC battery system? **A**
   - A. A constant 30 volt reading on the GMDSS console voltmeter.
   - B. After testing the station on battery power, the ammeter reading indicates a high rate of charge that then declines.
   - C. After testing the station on battery power, a voltmeter reading of 30 volts for brief period followed by a steady 26 volt reading.
   - D. None of the above.

5. Your antenna tuner becomes totally inoperative. What would you do to obtain operation on both the 8 MHz and 22 MHz frequency bands? **D**
   - A. Without an operating antenna tuner, transmission is impossible.
   - B. It is impossible to obtain operation on 2 different HF bands, without an operating antenna tuner.
   - C. Bypass the antenna tuner and shorten the whip to 15 ft.
   - D. Bypass the antenna tuner. Use a straight whip or wire antenna approximately 30 ft long.

6. Which of the following conditions would be a symptom of malfunction in a 2182 kHz radiotelephone system that must be reported to the Master, then logged appropriately. **B**
   - A. Much higher noise level observed during daytime operation.
   - B. No indication of power output when speaking into the microphone.
   - C. When testing a radiotelephone alarm on 2182 kHz into an artificial antenna, the Distress frequency watch receiver becomes unmuted, an improper testing procedure.
   - D. Failure to contact a shore station 600 nautical miles distant during daytime operation.

Subelement D - Other Equipment: 6 Key Topics - 6 Exam Questions

Key Topic 19: Antennas

1. What are the antenna requirements of a VHF telephony coast, maritime utility or ship station? **A**
   - A. The shore or on-board antenna must be vertically polarized.
   - B. The antenna array must be type-accepted for 30-200 MHz operation by the FCC.
   - C. The horizontally-polarized antenna must be positioned so as not to cause excessive interference to other stations.
   - D. The antenna must be capable of being energized by an output in excess of 100 watts.

2. What is the antenna requirement of a radiotelephone installation aboard a passenger vessel? **B**
   - A. The antenna must be located a minimum of 15 meters from the radiotelegraph antenna.
   - B. The antenna must be vertically polarized and as non-directional and efficient as is practicable for the transmission and reception of ground waves over seawater.
   - C. An emergency reserve antenna system must be provided for communications on 156.800 MHz.
   - D. All antennas must be tested and the operational results logged at least once during each voyage.

3. What is the most common type of antenna for GMDSS VHF? **D**
   - A. Horizontally polarized circular antenna.
   - B. Long wire antenna.
   - C. Both of the above.
   - D. None of the above.

4. What is the purpose of the antenna tuner? **A**
   - A. It alters the electrical characteristics of the antenna to match the frequency in use.
   - B. It physically alters the length of the antenna to match the frequency in use.
   - C. It makes the antenna look like a half-wave antenna at the frequency in use.
   - D. None of the above.

5. What advantage does a vertical whip have over a long wire? **B**
   - A. It radiates more signal fore and aft.
   - B. It radiates equally well in all directions.
   - C. It radiates a strong signal vertically.
   - D. None of the above.

6. A vertical whip antenna has a radiation pattern best described by? **C**
   - A. A figure eight.
   - B. A cardioid.
   - C. A circle.
   - D. An ellipse.

Key Topic 20: Power Sources

1. For a small passenger vessel inspection, reserve power batteries must be tested: **D**
   - A. At intervals not exceeding every 3 months.
   - B. At intervals not exceeding every 6 months
   - C. Before any new voyage
   - D. At intervals not exceeding 12 months, or during the inspection.

2. What are the characteristics of the Reserve Source of Energy under GMDSS? **C**
   - A. Supplies independent HF and MF installations at the same time.
   - B. Cannot be independent of the propelling power of the ship.
   - C. Must be independent of the ship's electrical system when the RSE is needed to supply power to the GMDSS equipment.
   - D. Must be incorporated into the ship's electrical system.

3. Which of the following terms is defined as a back-up power source that provides power to radio installations for the purpose of conducting Distress and Safety communications when the vessel's main and emergency generators cannot? **B**
   - A. Emergency Diesel Generator.
   - B. Reserve Source of Energy.
   - C. Reserve Source of Diesel Power.
   - D. Emergency Back-up Generator.

4. In the event of failure of the main and emergency sources of electrical power, what is the term for the source required to supply the GMDSS console with power for conducting distress and other radio communications? **C**
   - A. Emergency power.
   - B. Ship's emergency diesel generator.
   - C. Reserve source of energy.
   - D. Ship's standby generator

5. What is the requirement for emergency and reserve power in GMDSS radio installations? **D**
   - A. An emergency power source for radio communications is not required if a vessel has proper reserve power (batteries).
   - B. A reserve power source is not required for radio communications.
   - C. Only one of the above is required if a vessel is equipped with a second 406 EPIRB as a backup means of sending a distress alert.
   - D. All newly constructed ships under GMDSS must have both emergency and reserve power sources for radio communications.

6. What is the meaning of "Reserve Source of Energy"? **A**
   - A. The supply of electrical energy sufficient to operate the radio installations for the purpose of conducting Distress and Safety communications in the event of failure of the ship's main and emergency sources of electrical power.
   - B. High caloric value items for lifeboat, per SOLAS regulations.
   - C. Diesel fuel stored for the purpose of operating the powered survival craft for a period equal to or exceeding the U.S.C.G. and SOLAS requirements.
   - D. None of these.

Key Topic 21: EPIRBs

1. What is an EPIRB? **A**
   - A. A battery-operated emergency position-indicating radio beacon that floats free of a sinking ship.
   - B. An alerting device notifying mariners of imminent danger.
   - C. A satellite-based maritime distress and safety alerting system.
   - D. A high-efficiency audio amplifier.

2. When are EPIRB batteries changed? **B**
   - A. After emergency use; after battery life expires.
   - B. After emergency use or within the month and year replacement date printed on the EPIRB.
   - C. After emergency use; every 12 months when not used.
   - D. Whenever voltage drops to less than 20% of full charge.

3. If a ship sinks, what device is designed to float free of the mother ship, is turned on automatically and transmits a distress signal? **A**
   - A. An emergency position indicating radio beacon.
   - B. EPIRB on 2182 kHz and 405.025 kHz.
   - C. Bridge-to-bridge transmitter on 2182 kHz.
   - D. Auto alarm keyer on any frequency.

4. How do you cancel a false EPIRB distress alert? **C**
   - A. Transmit a DSC distress alert cancellation.
   - B. Transmit a broadcast message to "all stations" canceling the distress message.
   - C. Notify the Coast Guard or rescue coordination center at once.
   - D. Make a radiotelephony "distress cancellation" transmission on 2182 kHz.

5. What is the COSPAS-SARSAT system? **B**
   - A. A global satellite communications system for users in the maritime, land and aeronautical mobile services.
   - B. An international satellite-based search and rescue system.
   - C. A broadband military satellite communications network.
   - D. A Wide Area Geostationary Satellite program (WAGS).

6. What is an advantage of a 406 MHz satellite EPIRB? **D**
   - A. It is compatible with the COSPAS-SARSAT Satellites and Global Maritime Distress Safety System (GMDSS) regulations.
   - B. Provides a fast, accurate method for the Coast Guard to locating and rescuing persons in distress.
   - C. Includes a digitally encoded message containing the ship's identity and nationality.
   - D. All of the above.

Key Topic 22: SARTs

1. In which frequency band does a search and rescue transponder operate? **D**
   A. 3 GHz.
   B. S-band.
   C. 406 MHz.
   D. 9 GHz.

2. How should the signal from a Search And Rescue Radar Transponder appear on a RADAR display? **C**
   A. A series of dashes.
   B. A series of spirals all originating from the range and bearing of the SART.
   C. A series of 12 equally spaced dots.
   D. A series of twenty dashes.

3. What is the purpose of the SART's audible tone alarm? **A**
   A. It informs survivors that assistance may be nearby.
   B. It informs survivors when the battery's charge condition has weakened.
   C. It informs survivors when the SART switches to the "standby" mode.
   D. It informs survivors that a nearby vessel is signaling on DSC.

4. Which statement is true regarding the SART? **D**
   A. This is a performance monitor attached to at least one S-band navigational RADAR system.
   B. This is a 9 GHz transponder capable of being received by another vessel's S-band navigational RADAR system.
   C. This is a performance monitor attached to at least one X-band navigational RADAR system.
   D. This is a 9 GHz transponder capable of being received by vessel's X-band navigational RADAR system.

5. At what point does a SART begin transmitting? **C**
   A. It immediately begins radiating when placed in the "on" position.
   B. It must be manually activated.
   C. If it has been placed in the "on" position, it will respond when it has been interrogated by a 9-GHz RADAR signal.
   D. If it has been placed in the "on" position, it will begin transmitting immediately upon detecting that it is in water.

6. How can a SART's effective range be maximized? **B**
   A. The SART should be placed in water immediately upon activation.
   B. The SART should be held as high as possible.
   C. Switch the SART into the "high" power position.
   D. If possible, the SART should be mounted horizontally so that its signal matches that of the searching RADAR signal.

Key Topic 23: Survival Craft VHF

1. Which statement is NOT true regarding the requirements of survival craft portable two-way VHF radiotelephone equipment? **C**

   A. Watertight to a depth of 1 meter for 5 minutes.
   B. Effective radiated power should be a minimum of 0.25 watts.
   C. Operates simplex on Ch-70 and at least one other channel.
   D. The antenna is fixed and non-removable.

2. Which statement is NOT true regarding the requirements of survival craft portable two-way VHF radiotelephone equipment? **A**

   A. Operation on Ch-13.
   B. Effective radiated power should be a minimum of 0.25 Watts.
   C. Simplex voice communications only.
   D. Operation on Ch-16.

3. With what other stations may portable survival craft transceivers communicate? **D**

   A. Communication is permitted between survival craft.
   B. Communication is permitted between survival craft and ship.
   C. Communication is permitted between survival craft and rescue unit.
   D. All of the above.

4. Equipment for radiotelephony use in survival craft stations under GMDSS must have what capability? **A**

   A. Operation on Ch-16.
   B. Operation on 457.525 MHz.
   C. Operation on 121.5 MHz.
   D. Any one of these.

5. Equipment for radiotelephony use in survival craft stations under GMDSS must have what characteristic(s)? **D**

   A. Operation on Ch-16.
   B. Watertight.
   C. Permanently-affixed antenna.
   D. All of these.

6. What is the minimum power of the SCT? **B**

   A. Five watts.
   B. One watt.
   C. 1/4 watt.
   D. None of the above.

Key Topic 24: NAVTEX

1. NAVTEX broadcasts are sent: **B**

   A. Immediately following traffic lists.
   B. In categories of messages indicated by a single letter or identifier.
   C. On request of maritime mobile stations.
   D. Regularly, after the radiotelephone silent periods.

2. MSI can be obtained by one (or more) of the following: **D**

   A. NAVTEX.
   B. SafetyNET.
   C. HF NBDP.
   D. All of the above.

3. Which of the following is the primary frequency that is used exclusively for NAVTEX broadcasts internationally? **A**

   A. 518 kHz.
   B. 2187.5 kHz.
   C. 4209.5 kHz.
   D. VHF channel 16 when the vessel is sailing in Sea Area A1, and 2187.5 kHz when in Sea Area A2.

4. What means are used to prevent the reception of unwanted broadcasts by vessels utilizing the NAVTEX system? **C**

   A. Operating the receiver only during daytime hours.
   B. Coordinating reception with published broadcast schedules.
   C. Programming the receiver to reject unwanted broadcasts.
   D. Automatic receiver de-sensitization during night hours.

5. When do NAVTEX broadcasts typically achieve maximum transmitting range? **B**

   A. Local noontime.
   B. Middle of the night.
   C. Sunset.
   D. Post sunrise.

6. What is the transmitting range of most NAVTEX stations? **C**

   A. Typically 50-100 nautical miles (90-180 km) from shore.
   B. Typically upwards of 1000 nautical miles (1800 km) during the daytime.
   C. Typically 200-400 nautical miles (360-720 km).
   D. It is limited to line-of-sight or about 30 nautical miles (54 km).

General Radiotelephone
Element 3 Questions

Element 3: General Radiotelephone Operator License. 100 questions concerning electronic fundamentals and techniques required to adjust, repair and maintain radio transmitters and receivers at stations licensed by the FCC in the aviation, maritime and international fixed public radio services. By the way, the examiner will not grade your Element 3 exam until after you have successfully passed the Element 1 exam.

A General Radiotelephone Operator License (GROL) must be held by any person that adjusts, repairs and maintains aviation and marine radio transmitters and receivers, including operating certain maritime land radio stations, compulsory equipped ship radiotelephone stations.

The 100 questions asked on your exam will be taken for an FCC question pool of 600 questions covering 17 subelement question groups. Within these groups are a grand total of 100 "Key Topics".

Your Element 3 exam will consist of one randomly selected multiple-choice question from each of the 100 "Key Topics" for a complete examination. The minimum passing score is 75 of the multiple-choice questions answered correctly. Almost every page In Element 3 has a complete Key Topic of six questions and you will see one of these on your exam.

Contents of the Element 3 Examination:

Subelement A - Principles

Electrical Elements - Key Topic 1
Magnetism - Key topic 2
Materials - Key Topic 3
Resistance, Capacitance & Inductance - Key Topic 4
Semi-Conductors - Key Topic 5
Electrical Measurements - Key Topic 6
Waveforms - Key Topic 7
Conduction - Key Topic 8

Subelement B - Electrical Math

Ohm's Law-1 - Key Topic 9
Ohm's Law-2 - Key Topic 10
Frequency - Key Topic 11
Waveforms - Key Topic 12
Power Relationships - Key Topic 13
RC Time Constants-1 - Key Topic 14
RC Time Constants-2 - Key Topic 15
Impedance Networks-1 - Key Topic 16
Impedance Networks-2 - Key Topic 17
Calculations - Key Topic 18

Subelement C - Components

Photoconductive Devices - Key Topic 19
Capacitors - Key Topic 20
Transformers - Key Topic 21
Voltage Regulators, Zener Diodes - Key Topic 22
SCRs, Triacs - Key Topic 23
Diodes - Key Topic 24
Transistors-1 - Key Topic 25
Transistors-2 - Key Topic 26
Light Emitting Diodes - Key Topic 27
Devices - Key Topic 28

Subelement D - Circuits

R-L-C Circuits - Key Topic 29
Op Amps - Key Topic 30
Phase Locked Loops; Voltage Controlled Oscillators; Mixers - Key Topic 31
Schematics - Key Topic 32

Subelement E - Digital Logic

Types Of Logic - Key Topic 33
Logic Gates - Key Topic 34
Logic Levels - Key Topic 35
Flip- Flops - Key Topic 36
Multivibrators - Key Topic 37
Memory - Key Topic 38
Microprocessors - Key Topic 39
Counters, Dividers, Converters - Key Topic 40

Subelement F - Receivers

Subelement H - Modulation

Subelement I - Power Sources

Subelement J - Antennas

Subelement K - Aircraft

Subelement L - Installation, Maintenance and Repair

Indicating Meters - Key Topic 74
Test Equipment - Key Topic 75
Oscilloscopes - Key Topic 76
Specialized Instruments - Key Topic 77
Measurement Procedures - Key Topic 78
Repair Procedures - Key Topic 79
Installation Codes and Procedures - Key Topic 80
Troubleshooting - Key Topic 81

Subelement M - Communications Technology

Types of Transmissions - Key Topic 82
Coding and Multiplexing - Key Topic 83
Signal Processing, Software and Codes - Key Topic 84

Subelement N - Marine

VHF- Key Topic 85
MF - HF, SSB and Sitor - Key Topic 86
Survival Craft Equipment: VHF , SARTs & EPIRBs - Key Topic 87
FAX, NAVTEX - Key Topic 88
NMEA Data - Key Topic 89

Subelement O - Radar

RADAR Theory – Key Topic 90
Components - Key Topic 91
Range, Pulse Width and Repetition Rate - Key Topic 92
Antennas and Waveguides - Key Topic 93
RADAR Equipment - Key Topic 94

Subelement P - Satellite

Low Earth Orbit Systems - Key Topic 95
INMARSAT Communications Systems-1 - Key Topic 96
INMARSAT Communications Systems-2 - Key Topic 97
GPS - Key Topic 98

Sub-element Q - Safety

Radiation Exposure - Key Topic 99
Safety Steps - Key Topic 100

FCC Commercial Element 3

Subelement A – Principles: 8 Key Topics - 8 Exam Questions

Key Topic 1: Electrical Elements

1. The product of the readings of an AC voltmeter and AC ammeter is called: **A**

   A. Apparent power.
   B. True power.
   C. Power factor.
   D. Current power.

2. What is the basic unit of electrical power? **B**

   A. Ohm.
   B. Watt.
   C. Volt.
   D. Ampere.

3. What is the term used to express the amount of electrical energy stored in an electrostatic field? **A**

   A. Joules.
   B. Coulombs.
   C. Watts.
   D. Volts.

4. What device is used to store electrical energy in an electrostatic field? **C**

   A. Battery.
   B. Transformer.
   C. Capacitor.
   D. Inductor.

5. What formula would determine the inductive reactance of a coil if frequency and coil inductance are known? **B**

   A. $X_L = \pi f L$
   B. $X_L = 2\pi f L$
   C. $X_L = 1 / 2\pi f C$
   D. $X_L = 1 / R2+X2$

6. What is the term for the out-of-phase power associated with inductors and capacitors? **D**

   A. Effective power.
   B. True power.
   C. Peak envelope power.
   D. Reactive power.

Key Topic 2: Magnetism

1. What determines the strength of the magnetic field around a conductor? **D**
   - A. The resistance divided by the current.
   - B. The ratio of the current to the resistance.
   - C. The diameter of the conductor.
   - D. The amount of current.

2. What will produce a magnetic field? **C**
   - A. A DC source not connected to a circuit.
   - B. The presence of a voltage across a capacitor.
   - C. A current flowing through a conductor.
   - D. The force that drives current through a resistor.

3. When induced currents produce expanding magnetic fields around conductors in a direction that opposes the original magnetic field, this is known as: **A**
   - A. Lenz's law.
   - B. Gilbert's law.
   - C. Maxwell's law.
   - D. Norton's law.

4. The opposition to the creation of magnetic lines of force in a magnetic circuit is known as: **D**
   - A. Eddy currents.
   - B. Hysteresis.
   - C. Permeability.
   - D. Reluctance.

5. What is meant by the term "back EMF"? **C**
   - A. A current equal to the applied EMF.
   - B. An opposing EMF equal to R times C (RC) percent of the applied EMF.
   - C. A voltage that opposes the applied EMF.
   - D. A current that opposes the applied EMF.

6. Permeability is defined as: **B**
   - A. The magnetic field created by a conductor wound on a laminated core and carrying current.
   - B. The ratio of magnetic flux density in a substance to the magnetizing force that produces it.
   - C. Polarized molecular alignment in a ferromagnetic material while under the influence of a magnetizing force.
   - D. None of these.

Key Topic 3: Materials

1. What metal is usually employed as a sacrificial anode for corrosion control purposes? **C**

   A. Platinum bushing.
   B. Lead bar.
   C. Zinc bar.
   D. Brass rod.

2. What is the relative dielectric constant for air? **A**

   A. 1
   B. 2
   C. 4
   D. 0

3. Which metal object may be least affected by galvanic corrosion when submerged in seawater? **D**

   A. Aluminum outdrive.
   B. Bronze through-hull.
   C. Exposed lead keel.
   D. Stainless steel propeller shaft.

4. Skin effect is the phenomenon where: **A**

   A. RF current flows in a thin layer of the conductor, closer to the surface, as frequency increases.
   B. RF current flows in a thin layer of the conductor, closer to the surface, asfrequency decreases.
   C. Thermal effects on the surface of the conductor increase the impedance.
   D. Thermal effects on the surface of the conductor decrease the impedance.

5. Corrosion resulting from electric current flow between dissimilar metals is called: **D**

   A. Electrolysis.
   B. Stray current corrosion.
   C. Oxygen starvation corrosion.
   D. Galvanic corrosion.

6. Which of these will be most useful for insulation at UHF frequencies? **B**

   A. Rubber.
   B. Mica.
   C. Wax impregnated paper.
   D. Lead.

Key Topic 4: Resistance, Capacitance & Inductance

1. What formula would calculate the total inductance of inductors in series? **B**

   A. $L_T = L_1 / L_2$
   B. $L_T = L_1 + L_2$
   C. $L_T = 1 / L_1 + L_2$
   D. $L_T = 1 / L_1 \times L_2$

2. Good conductors with minimum resistance have what type of electrons? **D**

   A. Few free electrons.
   B. No electrons.
   C. Some free electrons.
   D. Many free electrons.

3. Which of the 4 groups of metals listed below are the best low-resistance conductors? **A**

   A. Gold, silver, and copper.
   B. Stainless steel, bronze, and lead.
   C. Iron, lead, and nickel.
   D. Bronze, zinc, and manganese.

4. What is the purpose of a bypass capacitor? **C**

   A. It increases the resonant frequency of the circuit.
   B. It removes direct current from the circuit by shunting DC to ground.
   C. It removes alternating current by providing a low impedance path to ground.
   D. It forms part of an impedance transforming circuit.

5. How would you calculate the total capacitance of three capacitors in parallel? **B**

   A. $C_T = C_1 + C_2 / C_1 - C_2 + C_3$.
   B. $C_T = C_1 + C_2 + C_3$.
   C. $C_T = C_1 + C_2 / C_1 \times C_2 + C_3$.
   D. $C_T = 1 / C_1 + 1 / C_2 + 1 / C_3$.

6. How might you reduce the inductance of an antenna coil? **C**

   A. Add additional turns.
   B. Add more core permeability.
   C. Reduce the number of turns.
   D. Compress the coil turns.

Key Topic 5: Semi-conductors

1. What are the two most commonly-used specifications for a junction diode? **D**

   A. Maximum forward current and capacitance.
   B. Maximum reverse current and PIV (peak inverse voltage).
   C. Maximum reverse current and capacitance.
   D. Maximum forward current and PIV (peak inverse voltage).

2. What limits the maximum forward current in a junction diode? **B**

   A. The peak inverse voltage (PIV).
   B. The junction temperature.
   C. The forward voltage.
   D. The back EMF.

3. MOSFETs are manufactured with THIS protective device built into their gate to protect the device from static charges and excessive voltages: **C**

   A. Schottky diode.
   B. Metal oxide varistor (MOV).
   C. Zener diode.
   D. Tunnel diode.

4. What are the two basic types of junction field-effect transistors? **A**

   A. N-channel and P-channel.
   B. High power and low power.
   C. MOSFET and GaAsFET.
   D. Silicon FET and germanium FET.

5. A common emitter amplifier has: **B**

   A. Lower input impedance than a common base.
   B. More voltage gain than a common collector.
   C. Less current gain than a common base.
   D. Less voltage gain than a common collector.

6. How does the input impedance of a field-effect transistor compare with that of a bipolar transistor? **A**

   A. An FET has high input impedance; a bipolar transistor has low input impedance.
   B. One cannot compare input impedance without first knowing the supply voltage.
   C. An FET has low input impedance; a bipolar transistor has high input impedance.
   D. The input impedance of FETs and bipolar transistors is the same.

Key Topic 6: Electrical Measurements

1. An AC ammeter indicates: **B**
   A. Effective (TRM) values of current.
   B. Effective (RMS) values of current.
   C. Peak values of current.
   D. Average values of current.

2. By what factor must the voltage of an AC circuit, as indicated on the scale of an AC voltmeter, be multiplied to obtain the peak voltage value? **C**
   A. 0.707
   B. 0.9
   C. 1.414
   D. 3.14

3. What is the RMS voltage at a common household electrical power outlet? **D**
   A. 331-V AC.
   B. 82.7-V AC.
   C. 165.5-V AC.
   D. 117-V AC.

4. What is the easiest voltage amplitude to measure by viewing a pure sine wave signal on an oscilloscope? **A**
   A. Peak-to-peak.
   B. RMS.
   C. Average.
   D. DC.

5. By what factor must the voltage measured in an AC circuit, as indicated on the scale of an AC voltmeter, be multiplied to obtain the average voltage value? **C**
   A. 0.707
   B. 1.414
   C. 0.9
   D. 3.14

6. What is the peak voltage at a common household electrical outlet? **D**
   A. 234 volts.
   B. 117 volts.
   C. 331 volts.
   D. 165.5 volts.

Key Topic 7: Waveforms

1. What is a sine wave? **B**
   - A. A constant-voltage, varying-current wave.
   - B. A wave whose amplitude at any given instant can be represented by the projection of a point on a wheel rotating at a uniform speed.
   - C. A wave following the laws of the trigonometric tangent function.
   - D. A wave whose polarity changes in a random manner.

2. How many degrees are there in one complete sine wave cycle? **D**
   - A. 90 degrees.
   - B. 270 degrees.
   - C. 180 degrees.
   - D. 360 degrees.

3. What type of wave is made up of sine waves of the fundamental frequency and all the odd harmonics? **A**
   - A. Square.
   - B. Sine.
   - C. Cosine.
   - D. Tangent.

4. What is the description of a square wave? **D**
   - A. A wave with only 300 degrees in one cycle.
   - B. A wave whose periodic function is always negative.
   - C. A wave whose periodic function is always positive.
   - D. A wave that abruptly changes back and forth between two voltage levels and stays at these levels for equal amounts of time.

5. What type of wave is made up of sine waves at the fundamental frequency and all the harmonics? **A**
   - A. Sawtooth wave.
   - B. Square wave.
   - C. Sine wave.
   - D. Cosine wave.

6. What type of wave is characterized by a rise time significantly faster than the fall time (or vice versa)? **C**
   - A. Cosine wave.
   - B. Square wave.
   - C. Sawtooth wave.
   - D. Sine wave.

Key Topic 8: Conduction

1. What is the term used to identify an AC voltage that would cause the same heating in a resistor as a corresponding value of DC voltage? **C**

   A. Cosine voltage.
   B. Power factor.
   C. Root mean square (RMS).
   D. Average voltage.

2. What happens to reactive power in a circuit that has both inductors and capacitors? **B**

   A. It is dissipated as heat in the circuit.
   B. It alternates between magnetic and electric fields and is not dissipated.
   C. It is dissipated as inductive and capacitive fields.
   D. It is dissipated as kinetic energy within the circuit.

3. Halving the cross-sectional area of a conductor will: **C**

   A. Not affect the resistance.
   B. Quarter the resistance.
   C. Double the resistance.
   D. Halve the resistance.

4. Which of the following groups is correct for listing common materials in order of descending conductivity? **A**

   A. Silver, copper, aluminum, iron, and lead.
   B. Lead, iron, silver, aluminum, and copper.
   C. Iron, silver, aluminum, copper, and silver.
   D. Silver, aluminum, iron, lead, and copper.

5. How do you compute true power (power dissipated in the circuit) in a circuit where AC voltage and current are out of phase? **D**

   A. Multiply RMS voltage times RMS current.
   B. Subtract apparent power from the power factor.
   C. Divide apparent power by the power factor.
   D. Multiply apparent power times the power factor.

6. Assuming a power source to have a fixed value of internal resistance, maximum power will be transferred to the load when: **B**

   A. The load impedance is greater than the source impedance.
   B. The load impedance equals the internal impedance of the source.
   C. The load impedance is less than the source impedance.
   D. The fixed values of internal impedance are not relative to the power source.

Key Topic 9: Ohm's Law-1

1. What value of series resistor would be needed to obtain a full scale deflection on a 50 microamp DC meter with an applied voltage of 200 volts DC? **A**
   - A. 4 megohms.
   - B. 2 megohms.
   - C. 400 kilohms.
   - D. 200 kilohms.

2. Which of the following Ohms Law formulas is incorrect? **B**
   - A. I = E / R
   - B. I = R / E
   - C. E = I x R
   - D. R = E / I

3. If a current of 2 amperes flows through a 50-ohm resistor, what is the voltage across the resistor? **D**
   - A. 25 volts.
   - B. 52 volts.
   - C. 200 volts.
   - D. 100 volts.

4. If a 100-ohm resistor is connected across 200 volts, what is the current through the resistor? **A**
   - A. 2 amperes.
   - B. 1 ampere.
   - C. 300 amperes.
   - D. 20,000 amperes.

5. If a current of 3 amperes flows through a resistor connected to 90 volts, what is the resistance? **B**
   - A. 3 ohms.
   - B. 30 ohms.
   - C. 93 ohms.
   - D. 270 ohms.

6. A relay coil has 500 ohms resistance, and operates on 125 mA. What value of resistance should be connected in series with it to operate from 110 V DC? **C**
   - A. 150 ohms.
   - B. 220 ohms.
   - C. 380 ohms.
   - D. 470 ohms.

Key Topic 10: Ohm's Law-2

1. What is the peak-to-peak RF voltage on the 50 ohm output of a 100 watt transmitter? **D**

A. 70 volts.
B. 100 volts.
C. 140 volts.
D. 200 volts.

2. What is the maximum DC or RMS voltage that may be connected across a 20 watt, 2000 ohm resistor? **C**

A. 10 volts.
B. 100 volts.
C. 200 volts.
D. 10,000 volts.

3. A 500-ohm, 2-watt resistor and a 1500-ohm, 1-watt resistor are connected in parallel. What is the maximum voltage that can be applied across the parallel circuit without exceeding wattage ratings? **B**

A. 22.4 volts.
B. 31.6 volts.
C. 38.7 volts.
D. 875 volts.

4. In Figure 3B1, what is the voltage drop across R1? **C**

A. 9 volts.
B. 7 volts.
C. 5 volts.
D. 3 volts.

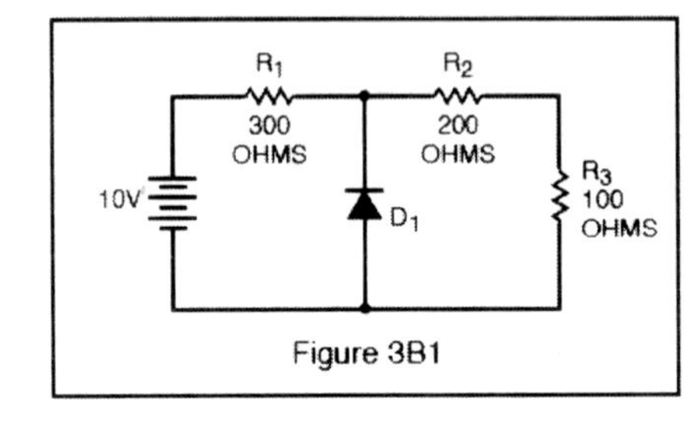

Figure 3B1

5. In Figure 3B2, what is the voltage drop across R1? **D**

A. 1.2 volts.
B. 2.4 volts.
C. 3.7 volts.
D. 9 volts.

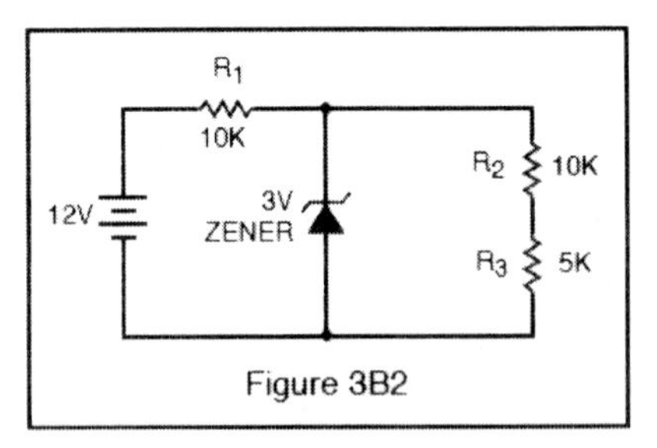

Figure 3B2

6. What is the maximum rated current-carrying capacity of a resistor marked "2000 ohms, 200 watts"? **A**

A. 0.316 amps.
B. 3.16 amps.
C. 10 amps.
D. 100 amps.

Key Topic 11: Frequency

1. What is the most the actual transmit frequency could differ from a reading of 462,100,000 Hertz on a frequency counter with a time base accuracy of ± 0.1 ppm? **A**
   - A. 46.21 Hz.
   - B. 0.1 MHz.
   - C. 462.1 Hz.
   - D. 0.2 MHz.

2. The second harmonic of a 380 kHz frequency is: **B**
   - A. 2 MHz.
   - B. 760 kHz.
   - C. 190 kHz.
   - D. 144.4 GHz.

3. What is the second harmonic of SSB frequency 4146 kHz? **A**
   - A. 8292 kHz.
   - B. 4.146 MHz.
   - C. 2073 kHz.
   - D. 12438 kHz.

4. What is the most the actual transmitter frequency could differ from a reading of 156,520,000 hertz on a frequency counter with a time base accuracy of ± 1.0 ppm? **C**
   - A. 165.2 Hz.
   - B. 15.652 kHz.
   - C. 156.52 Hz.
   - D. 1.4652 MHz.

5. What is the most the actual transmitter frequency could differ from a reading of 156,520,000 Hertz on a frequency counter with a time base accuracy of +/- 10 ppm? **B**
   - A. 146.52 Hz.
   - B. 1565.20 Hz.
   - C. 10 Hz.
   - D. 156.52 kHz.

6. What is the most the actual transmitter frequency could differ from a reading of 462,100,000 hertz on a frequency counter with a time base accuracy of ± 1.0 ppm? **D**
   - A. 46.21 MHz.
   - B. 10 Hz.
   - C. 1.0 MHz.
   - D. 462.1 Hz.

Key Topic 12: Waveforms

1. At pi/3 radians, what is the amplitude of a sine-wave having a peak value of 5 volts? **D**

   A. -4.3 volts.
   B. -2.5 volts.
   C. +2.5 volts.
   D. +4.3 volts.

2. At 150 degrees, what is the amplitude of a sine-wave having a peak value of 5 volts? **C**

   A. -4.3 volts.
   B. -2.5 volts.
   C. +2.5 volts.
   D. +4.3 volts.

3. At 240 degrees, what is the amplitude of a sine-wave having a peak value of 5 volts? **A**

   A. -4.3 volts.
   B. -2.5 volts.
   C. +2.5 volts.
   D. +4.3 volts.

4. What is the equivalent to the root-mean-square value of an AC voltage? **D**

   A. AC voltage is the square root of the average AC value.
   B. The DC voltage causing the same heating in a given resistor at the peak AC voltage.
   C. The AC voltage found by taking the square of the average value of the peak AC voltage.
   D. The DC voltage causing the same heating in a given resistor as the RMS AC voltage of the same value.

5. What is the RMS value of a 340-volt peak-to-peak pure sine wave? **C**

   A. 170 volts AC.
   B. 240 volts AC.
   C. 120 volts AC.
   D. 350 volts AC.

6. Determine the phase relationship between the two signals shown in Figure 3B3. **B**

   A. A is lagging B by 90 degrees.
   B. B is lagging A by 90 degrees.
   C. A is leading B by 180 degrees.
   D. B is leading A by 90 degrees.

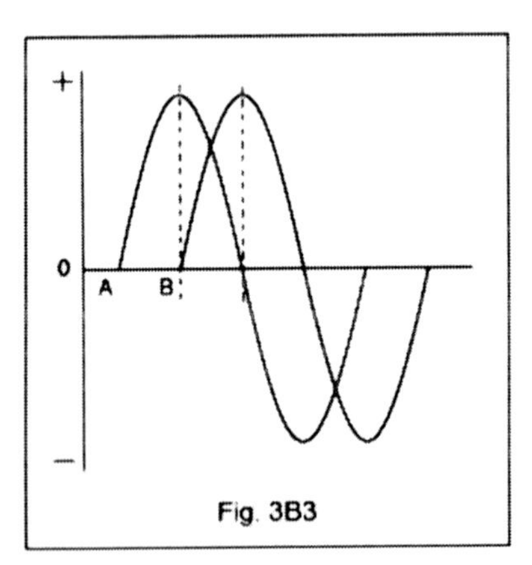

Fig. 3B3

Key Topic 13: Power Relationships

1. What does the power factor equal in an R-L circuit having a 60 degree phase angle between the voltage and the current? **C**

   A. 0.414
   B. 0.866
   C. 0.5
   D. 1.73

2. If a resistance to which a constant voltage is applied is halved, what power dissipation will result? **A**

   A. Double.
   B. Halved.
   C. Quadruple.
   D. Remain the same.

3. 746 watts, corresponding to the lifting of 550 pounds at the rate of one-foot-per-second, is the equivalent of how much horsepower? **D**

   A. One-quarter horsepower.
   B. One-half horsepower.
   C. Three-quarters horsepower.
   D. One horsepower.

4. In a circuit where the AC voltage and current are out of phase, how can the true power be determined? **A**

   A. By multiplying the apparent power times the power factor.
   B. By subtracting the apparent power from the power factor.
   C. By dividing the apparent power by the power factor.
   D. By multiplying the RMS voltage times the RMS current.

5. What does the power factor equal in an R-L circuit having a 45 degree phase angle between the voltage and the current? **D**

   A. 0.866
   B. 1.0
   C. 0.5
   D. 0.707

6. What does the power factor equal in an R-L circuit having a 30 degree phase angle between the voltage and the current? **B**

   A. 1.73
   B. 0.866
   C. 0.5
   D. 0.577

Key Topic 14: RC Time Constants-1

1. What is the term for the time required for the capacitor in an RC circuit to be charged to 63.2% of the supply voltage? **B**

   A. An exponential rate of one.
   B. One time constant.
   C. One exponential period.
   D. A time factor of one.

2. What is the meaning of the term "time constant of an RC circuit"? The time required to charge the capacitor in the circuit to: **D**

   A. 23.7% of the supply voltage.
   B. 36.8% of the supply voltage.
   C. 57.3% of the supply voltage.
   D. 63.2% of the supply voltage.

3. What is the term for the time required for the current in an RL circuit to build up to 63.2% of the maximum value? **A**

   A. One time constant.
   B. An exponential period of one.
   C. A time factor of one.
   D. One exponential rate.

4. What is the meaning of the term "time constant of an RL circuit"? The time required for the: **C**

   A. Current in the circuit to build up to 36.8% of the maximum value.
   B. Voltage in the circuit to build up to 63.2% of the maximum value.
   C. Current in the circuit to build up to 63.2% of the maximum value.
   D. Voltage in the circuit to build up to 36.8% of the maximum value.

5. After two time constants, the capacitor in an RC circuit is charged to what percentage of the supply voltage? **C**

   A. 36.8 %
   B. 63.2 %
   C. 86.5 %
   D. 95 %

6. After two time constants, the capacitor in an RC circuit is discharged to what percentage of the starting voltage? **B**

   A. 86.5 %
   B. 13.5 %
   C. 63.2 %
   D. 36.8 %

Key Topic 15: RC Time Constants-2

1. What is the time constant of a circuit having two 220-microfarad capacitors and two 1-megohm resistors all in parallel? **D**

   A. 22 seconds.
   B. 44 seconds.
   C. 440 seconds.
   D. 220 seconds.

2. What is the time constant of a circuit having two 100-microfarad capacitors and two 470-kilohm resistors all in series? **B**

   A. 470 seconds.
   B. 47 seconds.
   C. 4.7 seconds.
   D. 0.47 seconds.

3. What is the time constant of a circuit having a 100-microfarad capacitor and a 470-kilohm resistor in series? **C**

   A. 4700 seconds.
   B. 470 seconds.
   C. 47 seconds.
   D. 0.47 seconds.

4. What is the time constant of a circuit having a 220-microfarad capacitor and a 1-megohm resistor in parallel? **A**

   A. 220 seconds.
   B. 22 seconds.
   C. 2.2 seconds.
   D. 0.22 seconds.

5. What is the time constant of a circuit having two 100-microfarad capacitors and two 470-kilohm resistors all in parallel? **B**

   A. 470 seconds.
   B. 47 seconds.
   C. 4.7 seconds.
   D. 0.47 seconds.

6. What is the time constant of a circuit having two 220-microfarad capacitors and two 1-megohm resistors all in series? **A**

   A. 220 seconds.
   B. 55 seconds.
   C. 110 seconds.
   D. 440 seconds.

Key Topic 16: Impedance Networks-1

1. What is the impedance of a network composed of a 0.1-microhenry inductor in series with a 20-ohm resistor, at 30 MHz? Specify your answer in rectangular coordinates. **C**

   A. 20 -j19
   B. 19 +j20
   C. 20 +j19
   D. 19 -j20

2. In rectangular coordinates, what is the impedance of a network composed of a 0.1-microhenry inductor in series with a 30-ohm resistor, at 5 MHz? **D**

   A. 30 -j3
   B. 3 +j30
   C. 3 -j30
   D. 30 +j3

3. In rectangular coordinates, what is the impedance of a network composed of a 10-microhenry inductor in series with a 40-ohm resistor, at 500 MHz? **A**

   A. 40 +j31400
   B. 40 -j31400
   C. 31400 +j40
   D. 31400 -j40

4. In rectangular coordinates, what is the impedance of a network composed of a 1.0-millihenry inductor in series with a 200-ohm resistor, at 30 kHz? **B**

   A. 200 - j188
   B. 200 + j188
   C. 188 + j200
   D. 188 - j200

5. In rectangular coordinates, what is the impedance of a network composed of a 0.01-microfarad capacitor in parallel with a 300-ohm resistor, at 50 kHz? **C**

   A. 150 - j159
   B. 150 + j159
   C. 159 - j150
   D. 159 + j150

6. In rectangular coordinates, what is the impedance of a network composed of a 0.001-microfarad capacitor in series with a 400-ohm resistor, at 500 kHz? **D**

   A. 318 - j400
   B. 400 + j318
   C. 318 + j400
   D. 400 - j318

Key Topic 17: Impedance Networks-2

1. What is the impedance of a network composed of a 100-picofarad capacitor in parallel with a 4000-ohm resistor, at 500 KHz? Specify your answer in polar coordinates. **D**

   A. 2490 ohms, /51.5 degrees
   B. 4000 ohms, /38.5 degrees
   C. 5112 ohms, /-38.5 degrees
   D. 2490 ohms, /-51.5 degrees

2. In polar coordinates, what is the impedance of a network composed of a 100-ohm-reactance inductor in series with a 100-ohm resistor? **B**

   A. 121 ohms, /35 degrees
   B. 141 ohms, /45 degrees
   C. 161 ohms, /55 degrees
   D. 181 ohms, /65 degrees

3. In polar coordinates, what is the impedance of a network composed of a 400-ohm-reactance capacitor in series with a 300-ohm resistor? **C**

   A. 240 ohms, /36.9 degrees
   B. 240 ohms, /-36.9 degrees
   C. 500 ohms, /-53.1 degrees
   D. 500 ohms, /53.1 degrees

4. In polar coordinates, what is the impedance of a network composed of a 300-ohm-reactance capacitor, a 600-ohm-reactance inductor, and a 400-ohm resistor, all connected in series? **A**

   A. 500 ohms, /37 degrees
   B. 400 ohms, /27 degrees
   C. 300 ohms, /17 degrees
   D. 200 ohms, /10 degrees

5. In polar coordinates, what is the impedance of a network comprised of a 400-ohm-reactance inductor in parallel with a 300-ohm resistor? **B**

   A. 240 ohms, /-36.9 degrees
   B. 240 ohms, /36.9 degrees
   C. 500 ohms, /53.1 degrees
   D. 500 ohms, /-53.1 degrees

6. Using the polar coordinate system, what visual representation would you get of a voltage in a sinewave circuit? **D**

   A. To show the reactance which is present.
   B. To graphically represent the AC and DC component.
   C. To display the data on an XY chart.
   D. The plot shows the magnitude and phase angle.

Key Topic 18: Calculations

1. What is the magnitude of the impedance of a series AC circuit having a resistance of 6 ohms, an inductive reactance of 17 ohms, and zero capacitive reactance? **C**

   A. 6.6 ohms.
   B. 11 ohms.
   C. 18 ohms.
   D. 23 ohms.

2. A 1-watt, 10-volt Zener diode with the following characteristics: $I_{min.}$ = 5 mA; $I_{max.}$ = 95 mA; and Z = 8 ohms, is to be used as part of a voltage regulator in a 20-V power supply. Approximately what size current-limiting resistor would be used to set its bias to the midpoint of its operating range? **B**

   A. 100 ohms.
   B. 200 ohms.
   C. 1 kilohms.
   D. 2 kilohms.

3. Given a power supply with a no load voltage of 12 volts and a full load voltage of 10 volts, what is the percentage of voltage regulation? **C**

   A. 17 %
   B. 80 %
   C. 20 %
   D. 83 %

4. What turns ratio does a transformer need in order to match a source impedance of 500 ohms to a load of 10 ohms? **A**

   A. 7.1 to 1.
   B. 14.2 to 1.
   C. 50 to 1.
   D. None of these.

5. Given a power supply with a full load voltage of 200 volts and a regulation of 25%, what is the no load voltage? **D**

   A. 150 volts.
   B. 160 volts.
   C. 240 volts.
   D. 250 volts.

6. What is the conductance (G) of a circuit if 6 amperes of current flows when 12 volts DC is applied? **B**

   A. 0.25 Siemens (mhos).
   B. 0.50 Siemens (mhos).
   C. 1.00 Siemens (mhos).
   D. 1.25 Siemens (mhos).

Subelement C – Components: 10 Key Topics - 10 Exam Questions - 2 Drawings

Key Topic 19: Photoconductive Devices

1. What happens to the conductivity of photoconductive material when light shines on it? **A**

   A. It increases.
   B. It decreases.
   C. It stays the same.
   D. It becomes temperature dependent.

2. What is the photoconductive effect? **B**

   A. The conversion of photon energy to electromotive energy.
   B. The increased conductivity of an illuminated semiconductor junction.
   C. The conversion of electromotive energy to photon energy.
   D. The decreased conductivity of an illuminated semiconductor junction.

3. What does the photoconductive effect in crystalline solids produce a noticeable change in? **D**

   A. The capacitance of the solid.
   B. The inductance of the solid.
   C. The specific gravity of the solid.
   D. The resistance of the solid.

4. What is the description of an optoisolator? **A**

   A. An LED and a photosensitive device.
   B. A P-N junction that develops an excess positive charge when exposed to light.
   C. An LED and a capacitor.
   D. An LED and a lithium battery cell.

5. What happens to the conductivity of a photosensitive semiconductor junction when it is illuminated? **B**

   A. The junction resistance is unchanged.
   B. The junction resistance decreases.
   C. The junction resistance becomes temperature dependent.
   D. The junction resistance increases

6. What is the description of an optocoupler? **C**

   A. A resistor and a capacitor.
   B. Two light sources modulated onto a mirrored surface.
   C. An LED and a photosensitive device.
   D. An amplitude modulated beam encoder.

Key Topic 20: Capacitors

1. What factors determine the capacitance of a capacitor? **D**

   A. Voltage on the plates and distance between the plates.
   B. Voltage on the plates and the dielectric constant of the material between the plates.
   C. Amount of charge on the plates and the dielectric constant of the material between the plates.
   D. Distance between the plates and the dielectric constant of the material between the plates.

2. In Figure 3C4, if a small variable capacitor were installed in place of the dashed line, it would? **C**

   A. Increase gain.
   B. Increase parasitic oscillations.
   C. Decrease parasitic oscillations.
   D. Decrease crosstalk.

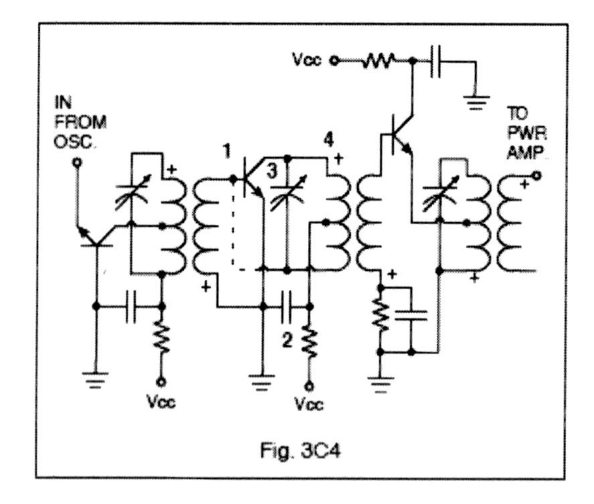

Fig. 3C4

3. In Figure 3C4, which component (labeled 1 through 4) is used to provide a signal ground? **B**

   A. 1
   B. 2
   C. 3
   D. 4

4. In Figure 3C5, which capacitor (labeled 1 through 4) is being used as a bypass capacitor? **C**

   A. 1
   B. 2
   C. 3
   D. 4

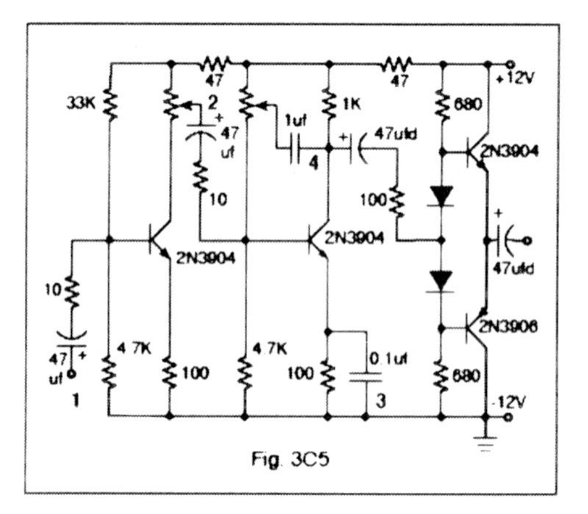

Fig. 3C5

5. In Figure 3C5, the 1 μF capacitor is connected to a potentiometer that is used to: **D**

   A. Increase gain.
   B. Neutralize amplifier.
   C. Couple.
   D. Adjust tone.

6. What is the purpose of a coupling capacitor? **A**

   A. It blocks direct current and passes alternating current.
   B. It blocks alternating current and passes direct current.
   C. It increases the resonant frequency of the circuit.
   D. It decreases the resonant frequency of the circuit.

Key Topic 21: Transformers

1. A capacitor is sometimes placed in series with the primary of a power transformer to: **A**
   A. Improve the power factor.
   B. Improve output voltage regulation.
   C. Rectify the primary windings.
   D. None of these.

2. A transformer used to step up its input voltage must have: **B**
   A. More turns of wire on its primary than on its secondary.
   B. More turns of wire on its secondary than on its primary.
   C. Equal number of primary and secondary turns of wire.
   D. None of the above statements are correct.

3. A transformer primary of 2250 turns connected to 120 VAC will develop what voltage across a 500-turn secondary? **A**
   A. 26.7 volts.
   B. 2300 volts.
   C. 1500 volts.
   D. 5.9 volts.

4. What is the ratio of the output frequency to the input frequency of a single-phase full-wave rectifier? **C**
   A. 1:1.
   B. 1:2.
   C. 2:1.
   D. None of these.

5. A power transformer has a single primary winding and three secondary windings producing 5.0 volts, 12.6 volts, and 150 volts. Assuming similar wire sizes, which of the three secondary windings will have the highest measured DC resistance? **B**
   A. The 12.6 volt winding.
   B. The 150 volt winding.
   C. The 5.0 volt winding.
   D. All will have equal resistance values.

6. A power transformer has a primary winding of 200 turns of #24 wire and a secondary winding consisting of 500 turns of the same size wire. When 20 volts are applied to the primary winding, the expected secondary voltage will be: **D**
   A. 500 volts.
   B. 25 volts.
   C. 10 volts.
   D. 50 volts.

Key Topic 22: Voltage Regulators, Zener Diodes

1. In a linear electronic voltage regulator: **D**
   - A. The output is a ramp voltage.
   - B. The pass transistor switches from the "off" state to the "on" state.
   - C. The control device is switched on or off, with the duty cycle proportional to the line or load conditions.
   - D. The conduction of a control element is varied in direct proportion to the line voltage or load current.

2. A switching electronic voltage regulator: **C**
   - A. Varies the conduction of a control element in direct proportion to the line voltage or load current.
   - B. Provides more than one output voltage.
   - C. Switches the control device on or off, with the duty cycle proportional to the line or load conditions.
   - D. Gives a ramp voltage at its output.

3. What device is usually used as a stable reference voltage in a linear voltage regulator? **A**
   - A. Zener diode.
   - B. Tunnel diode.
   - C. SCR.
   - D. Varactor diode.

4. In a regulated power supply, what type of component will most likely be used to establish a reference voltage? **D**
   - A. Tunnel Diode.
   - B. Battery.
   - C. Pass Transistor.
   - D. Zener Diode.

5. A three-terminal regulator: **C**
   - A. Supplies three voltages with variable current.
   - B. Supplies three voltages at a constant current.
   - C. Contains a voltage reference, error amplifier, sensing resistors and transistors, and a pass element.
   - D. Contains three error amplifiers and sensing transistors.

6. What is the range of voltage ratings available in Zener diodes? **B**
   - A. 1.2 volts to 7 volts.
   - B. 2.4 volts to 200 volts and above.
   - C. 3 volts to 2000 volts.
   - D. 1.2 volts to 5.6 volts.

Key Topic 23: SCRs, Triacs

1. How might two similar SCRs be connected to safely distribute the power load of a circuit? **C**

   A. In series.
   B. In parallel, same polarity.
   C. In parallel, reverse polarity.
   D. In a combination series and parallel configuration.

2. What are the three terminals of an SCR? **A**

   A. Anode, cathode, and gate.
   B. Gate, source, and sink.
   C. Base, collector, and emitter.
   D. Gate, base 1, and base 2.

3. Which of the following devices acts as two SCRs connected back to back, but facing in opposite directions and sharing a common gate? **D**

   A. JFET.
   B. Dual-gate MOSFET.
   C. DIAC.
   D. TRIAC.

4. What is the transistor called that is fabricated as two complementary SCRs in parallel with a common gate terminal? **A**

   A. TRIAC.
   B. Bilateral SCR.
   C. Unijunction transistor.
   D. Field effect transistor.

5. What are the three terminals of a TRIAC? **D**

   A. Emitter, base 1, and base 2.
   B. Base, emitter, and collector.
   C. Gate, source, and sink.
   D. Gate, anode 1, and anode 2.

6. What circuit might contain a SCR? **B**

   A. Filament circuit of a tube radio receiver.
   B. A light-dimming circuit.
   C. Shunt across a transformer primary.
   D. Bypass capacitor circuit to ground.

Key Topic 24: Diodes

1. What is one common use for PIN diodes? **B**
   A. Constant current source.
   B. RF switch.
   C. Constant voltage source.
   D. RF rectifier.

2. What is a common use of a hot-carrier diode? **D**
   A. Balanced inputs in SSB generation.
   B. Variable capacitance in an automatic frequency control circuit.
   C. Constant voltage reference in a power supply.
   D. VHF and UHF mixers and detectors.

3. Structurally, what are the two main categories of semiconductor diodes? **A**
   A. Junction and point contact.
   B. Electrolytic and junction.
   C. Electrolytic and point contact.
   D. Vacuum and point contact.

4. What special type of diode is capable of both amplification and oscillation? **C**
   A. Zener diodes.
   B. Point contact diodes.
   C. Tunnel diodes.
   D. Junction diodes.

5. What type of semiconductor diode varies its internal capacitance as the voltage applied to its terminals varies? **B**
   A. Tunnel diode.
   B. Varactor diode.
   C. Silicon-controlled rectifier.
   D. Zener diode.

6. What is the principal characteristic of a tunnel diode? **C**
   A. High forward resistance.
   B. Very high PIV(peak inverse voltage).
   C. Negative resistance region.
   D. High forward current rating.

Key Topic 25: Transistors-1

1. What is the meaning of the term "alpha" with regard to bipolar transistors? The change of: **D**

   A. Collector current with respect to base current.
   B. Base current with respect to collector current.
   C. Collector current with respect to gate current.
   D. Collector current with respect to emitter current.

2. What are the three terminals of a bipolar transistor? **B**

   A. Cathode, plate and grid.
   B. Base, collector and emitter.
   C. Gate, source and sink.
   D. Input, output and ground.

3. What is the meaning of the term "beta" with regard to bipolar transistors? The change of: **C**

   A. Base current with respect to emitter current.
   B. Collector current with respect to emitter current.
   C. Collector current with respect to base current.
   D. Base current with respect to gate current.

4. What are the elements of a unijunction transistor? **A**

   A. Base 1, base 2, and emitter.
   B. Gate, cathode, and anode.
   C. Gate, base 1, and base 2.
   D. Gate, source, and sink.

5. The beta cutoff frequency of a bipolar transistor is the frequency at which: **B**

   A. Base current gain has increased to 0.707 of maximum.
   B. Emitter current gain has decreased to 0.707 of maximum.
   C. Collector current gain has decreased to 0.707.
   D. Gate current gain has decreased to 0.707.

6. What does it mean for a transistor to be fully saturated? **A**

   A. The collector current is at its maximum value.
   B. The collector current is at its minimum value.
   C. The transistor's Alpha is at its maximum value.
   D. The transistor's Beta is at its maximum value.

Key Topic 26: Transistors-2

1. A common base amplifier has: **B**
   A. More current gain than common emitter or common collector.
   B. More voltage gain than common emitter or common collector.
   C. More power gain than common emitter or common collector.
   D. Highest input impedance of the three amplifier configurations.

2. What does it mean for a transistor to be cut off? **C**
   A. There is no base current.
   B. The transistor is at its Class A operating point.
   C. There is no current between emitter and collector.
   D. There is maximum current between emitter and collector.

3. An emitter-follower amplifier has: **D**
   A. More voltage gain than common emitter or common base.
   B. More power gain than common emitter or common base.
   C. Lowest input impedance of the three amplifier configurations.
   D. More current gain than common emitter or common base.

4. What conditions exists when a transistor is operating in saturation? **A**
   A. The base-emitter junction and collector-base junction are both forward biased.
   B. The base-emitter junction and collector-base junction are both reverse biased.
   C. The base-emitter junction is reverse biased and the collector-base junction is forward biased.
   D. The base-emitter junction is forward biased and the collector-base junction is reverse biased.

5. For current to flow in an NPN silicon transistor's emitter-collector junction, the base must be: **C**
   A. At least 0.4 volts positive with respect to the emitter.
   B. At a negative voltage with respect to the emitter.
   C. At least 0.7 volts positive with respect to the emitter.
   D. At least 0.7 volts negative with respect to the emitter.

6. When an NPN transistor is operating as a Class A amplifier, the base-emitter junction: **D**
   A. And collector-base junction are both forward biased.
   B. And collector-base junction are both reverse biased.
   C. Is reverse biased and the collector-base junction is forward biased.
   D. Is forward biased and the collector-base junction is reverse biased.

Key Topic 27: Light Emitting Diodes

1. What type of bias is required for an LED to produce luminescence? **B**
   - A. Reverse bias.
   - B. Forward bias.
   - C. Logic 0 (Lo) bias.
   - D. Logic 1 (Hi) bias.

2. What determines the visible color radiated by an LED junction? **D**
   - A. The color of a lens in an eyepiece.
   - B. The amount of voltage across the device.
   - C. The amount of current through the device.
   - D. The materials used to construct the device.

3. What is the approximate operating current of a light-emitting diode? **A**
   - A. 20 mA.
   - B. 5 mA.
   - C. 10 mA.
   - D. 400 mA.

4. What would be the maximum current to safely illuminate a LED? **D**
   - A. 1 amp.
   - B. 1 microamp.
   - C. 500 milliamps.
   - D. 20 mA.

5. An LED facing a photodiode in a light-tight enclosure is commonly known as a/an: **A**
   - A. Optoisolator.
   - B. Seven segment LED.
   - C. Optointerrupter.
   - D. Infra-red (IR) detector.

6. What circuit component must be connected in series to protect an LED? **C**
   - A. Bypass capacitor to ground.
   - B. Electrolytic capacitor.
   - C. Series resistor.
   - D. Shunt coil in series.

Key Topic 28: Devices

1. What describes a diode junction that is forward biased? **C**

   A. It is a high impedance.
   B. It conducts very little current.
   C. It is a low impedance.
   D. It is an open circuit.

2. Why are special precautions necessary in handling FET and CMOS devices? **B**

   A. They have fragile leads that may break off.
   B. They are susceptible to damage from static charges.
   C. They have micro-welded semiconductor junctions that are susceptible to breakage.
   D. They are light sensitive.

3. What do the initials CMOS stand for? **C**

   A. Common mode oscillating system.
   B. Complementary mica-oxide silicon.
   C. Complementary metal-oxide semiconductor.
   D. Complementary metal-oxide substrate.

4. What is the piezoelectric effect? **A**

   A. Mechanical vibration of a crystal by the application of a voltage.
   B. Mechanical deformation of a crystal by the application of a magnetic field.
   C. The generation of electrical energy by the application of light.
   D. Reversed conduction states when a P-N junction is exposed to light.

5. An electrical relay is a: **D**

   A. Current limiting device.
   B. Device used for supplying 3 or more voltages to a circuit.
   C. Component used mainly with HF audio amplifiers.
   D. Remotely controlled switching device.

6. In which oscillator circuit would you find a quartz crystal? **B**

   A. Hartley.
   B. Pierce
   C. Colpitts.
   D. All of the above.

Subelement D – Circuits: 4 Key Topics - 4 Exam Questions

Key Topic 29: R-L-C Circuits

1. What is the approximate magnitude of the impedance of a parallel R-L-C circuit at resonance? **A**

   A. Approximately equal to the circuit resistance.
   B. Approximately equal to X_L.
   C. Low, as compared to the circuit resistance.
   D. Approximately equal to X_C.

2. What is the approximate magnitude of the impedance of a series R-L-C circuit at resonance? **B**

   A. High, as compared to the circuit resistance.
   B. Approximately equal to the circuit resistance.
   C. Approximately equal to X_L.
   D. Approximately equal to X_C.

3. How could voltage be greater across reactances in series than the applied voltage? **D**

   A. Resistance.
   B. Conductance.
   C. Capacitance.
   D. Resonance.

4. What is the characteristic of the current flow in a series R-L-C circuit at resonance? **A**

   A. Maximum.
   B. Minimum.
   C. DC.
   D. Zero.

5. What is the characteristic of the current flow within the parallel elements in a parallel R-L-C circuit at resonance? **B**

   A. Minimum.
   B. Maximum.
   C. DC.
   D. Zero.

6. What is the relationship between current through a resonant circuit and the voltage across the circuit? **C**

   A. The current and voltage are 180 degrees out of phase.
   B. The current leads the voltage by 90 degrees.
   C. The voltage and current are in phase.
   D. The voltage leads the current by 90 degrees.

Key Topic 30: Op Amps

1. What is the main advantage of using an op-amp audio filter over a passive LC audio filter? **D**
   - A. Op-amps are largely immune to vibration and temperature change.
   - B. Most LC filter manufacturers have retooled to make op-amp filters.
   - C. Op-amps are readily available in a wide variety of operational voltages and frequency ranges.
   - D. Op-amps exhibit gain rather than insertion loss.

2. What are the characteristics of an inverting operational amplifier (op-amp) circuit? **C**
   - A. It has input and output signals in phase.
   - B. Input and output signals are 90 degrees out of phase.
   - C. It has input and output signals 180 degrees out of phase.
   - D. Input impedance is low while the output impedance is high.

3. Gain of a closed-loop op-amp circuit is determined by? **B**
   - A. The maximum operating frequency divided by the square root of the load impedance.
   - B. The op-amp's external feedback network.
   - C. Supply voltage and slew rate.
   - D. The op-amp's internal feedback network.

4. Where is the external feedback network connected to control the gain of a closed-loop op-amp circuit? **C**
   - A. Between the differential inputs.
   - B. From output to the non-inverting input.
   - C. From output to the inverting input.
   - D. Between the output and the differential inputs.

5. Which of the following op-amp circuits is operated open-loop? **D**
   - A. Non-inverting amp.
   - B. Inverting amp.
   - C. Active filter.
   - D. Comparator.

6. In the op-amp oscillator circuit shown in Figure 3D6, what would be the most noticeable effect if the capacitance of C were suddenly doubled? **A**
   - A. Frequency would be lower.
   - B. Frequency would be higher.
   - C. There would be no change. The inputs are reversed, therefore the circuit cannot function.
   - D. None of the above.

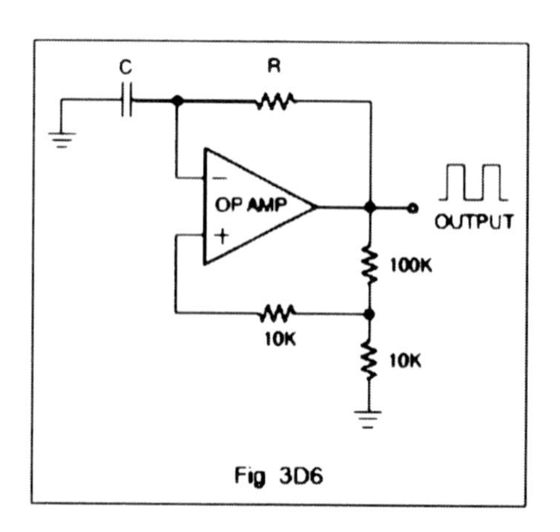

Fig 3D6

Key Topic 31: Phase Locked Loops (PLLs); Voltage Controlled Oscillators (VCOs); Mixers

1. What frequency synthesizer circuit uses a phase comparator, look-up table, digital-to-analog converter, and a low-pass antialias filter? **A**

   A. A direct digital synthesizer.
   B. Phase-locked-loop synthesizer.
   C. A diode-switching matrix synthesizer.
   D. A hybrid synthesizer.

2. A circuit that compares the output of a voltage-controlled oscillator (VCO) to a frequency standard and produces an error voltage that is then used to adjust the capacitance of a varactor diode used to control frequency in that same VCO is called what? **B**

   A. Doubly balanced mixer.
   B. Phase-locked loop.
   C. Differential voltage amplifier.
   D. Variable frequency oscillator.

3. RF input to a mixer is 200 MHz and the local oscillator frequency is 150 MHz. What output would you expect to see at the IF output prior to any filtering? **A**

   A. 50, 150, 200 and 350 MHz.
   B. 50 MHz.
   C. 350 MHz.
   D. 50 and 350 MHz.

4. What spectral impurity components might be generated by a phase-locked-loop synthesizer? **C**

   A. Spurs at discrete frequencies.
   B. Random spurs which gradually drift up in frequency.
   C. Broadband noise.
   D. Digital conversion noise.

5. In a direct digital synthesizer, what are the unwanted components on its output? **B**

   A. Broadband noise.
   B. Spurs at discrete frequencies.
   C. Digital conversion noise.
   D. Nyquist limit noise pulses.

6. What is the definition of a phase-locked loop (PLL) circuit? **D**

   A. A servo loop consisting of a ratio detector, reactance modulator, and voltage-controlled oscillator.
   B. A circuit also known as a monostable multivibrator.
   C. A circuit consisting of a precision push-pull amplifier with a differential input.
   D. A servo loop consisting of a phase detector, a low-pass filter and voltage-controlled oscillator.

Key Topic 32: Schematics

1. Given the combined DC input voltages, what would the output voltage be in the circuit shown in Figure 3D7? **D**

   A. 150 mV
   B. 5.5 V
   C. -15 mv
   D. -5.5 V

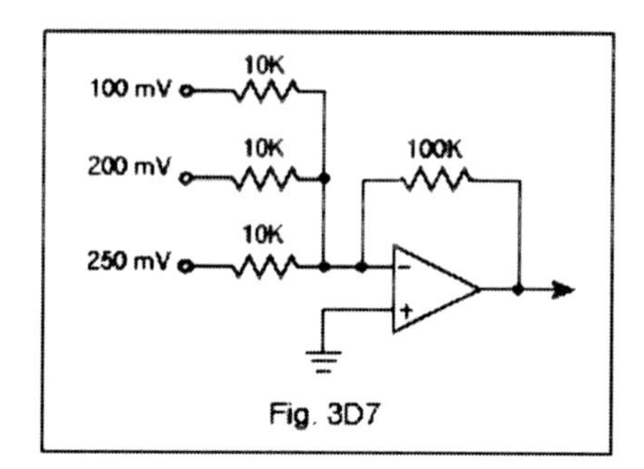

Fig. 3D7

2. Which lamps would be lit in the circuit shown in Figure 3D8? **C**

   A. 2, 3, 4, 5 and 6.
   B. 5, 6, 8 and 9.
   C. 2, 3, 4, 7 and 8.
   D. 1, 3, 5, 7 and 8.

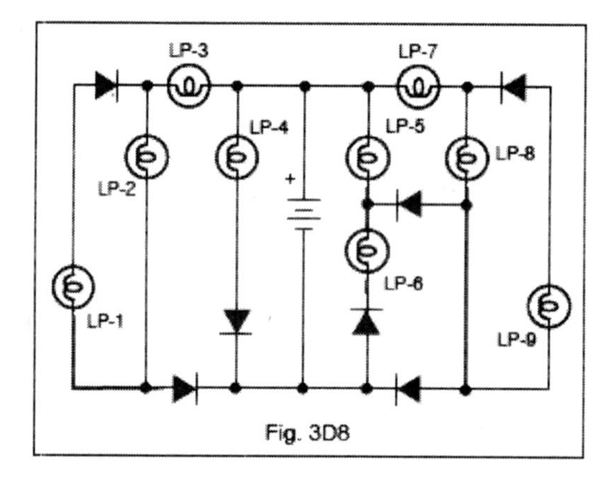

Fig. 3D8

3. What will occur if an amplifier input signal coupling capacitor fails open? **A**

   A. No amplification will occur, with DC within the circuit measuring normal.
   B. Improper biasing will occur within the amplifier stage.
   C. Oscillation and thermal runaway may occur.
   D. An AC hum will appear on the circuit output.

4. In Figure 3D9, determine if there is a problem with this regulated power supply and identify the problem. **D**

   A. R1 value is too low which would cause excessive base current and instantly destroy TR 1.
   B. D1 and D2 are reversed. The power supply simply would not function.
   C. TR1 is shown as an NPN and must be changed to a PNP.
   D. There is no problem with the circuit.

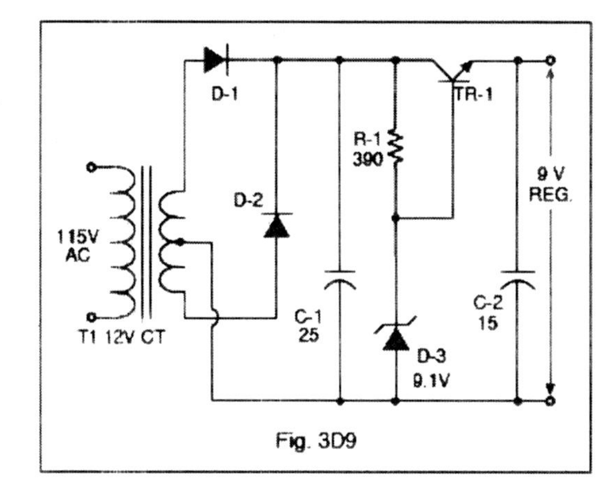

Fig. 3D9

5. In Figure 3D10 with a square wave input what would be the output? **C**

   A. 1
   B. 2
   C. 3
   D. 4

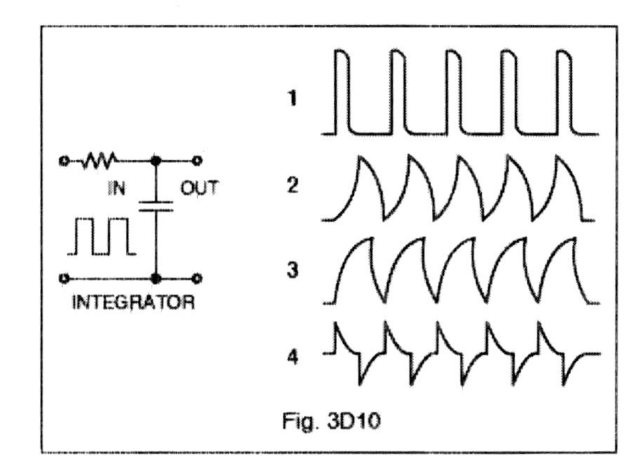

Fig. 3D10

6. With a pure AC signal input to the circuit shown in Figure 3D11, what output wave form would you expect to see on an oscilloscope display? **B**

   A. 1
   B. 2
   C. 3
   D. 4

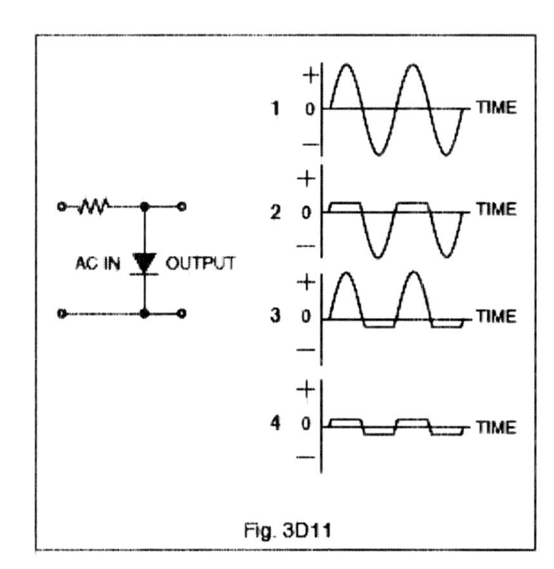

Fig. 3D11

Subelement E – Digital Logic: 8 Key Topics - 8 Exam Questions

Key Topic 33: Types of Logic

1. What is the voltage range considered to be valid logic low input in a TTL device operating at 5 volts? **C**

   A. 2.0 to 5.5 volts.
   B. -2.0 to -5.5 volts.
   C. Zero to 0.8 volts.
   D. 5.2 to 34.8 volts.

2. What is the voltage range considered to be a valid logic high input in a TTL device operating at 5.0 volts? **A**

   A. 2.0 to 5.5 volts.
   B. 1.5 to 3.0 volts.
   C. 1.0 to 1.5 volts.
   D. 5.2 to 34.8 volts.

3. What is the common power supply voltage for TTL series integrated circuits? **D**

   A. 12 volts.
   B. 13.6 volts.
   C. 1 volt.
   D. 5 volts.

4. TTL inputs left open develop what logic state? **A**

   A. A high-logic state.
   B. A low-logic state.
   C. Open inputs on a TTL device are ignored.
   D. Random high- and low-logic states.

5. Which of the following instruments would be best for checking a TTL logic circuit? **D**

   A. VOM.
   B. DMM.
   C. Continuity tester.
   D. Logic probe.

6. What do the initials TTL stand for? **B**

   A. Resistor-transistor logic.
   B. Transistor-transistor logic.
   C. Diode-transistor logic.
   D. Emitter-coupled logic.

Key Topic 34: Logic Gates

1. What is a characteristic of an AND gate? **B**

   A. Produces a logic "0" at its output only if all inputs are logic "1".
   B. Produces a logic "1" at its output only if all inputs are logic "1".
   C. Produces a logic "1" at its output if only one input is a logic "1".
   D. Produces a logic "1" at its output if all inputs are logic "0".

2. What is a characteristic of a NAND gate? **D**

   A. Produces a logic "0" at its output only when all inputs are logic "0".
   B. Produces a logic "1" at its output only when all inputs are logic "1".
   C. Produces a logic "0" at its output if some but not all of its inputs are logic "1".
   D. Produces a logic "0" at its output only when all inputs are logic "1".

3. What is a characteristic of an OR gate? **A**

   A. Produces a logic "1" at its output if any input is logic "1".
   B. Produces a logic "0" at its output if any input is logic "1".
   C. Produces a logic "0" at its output if all inputs are logic "1".
   D. Produces a logic "1" at its output if all inputs are logic "0".

4. What is a characteristic of a NOR gate? **C**

   A. Produces a logic "0" at its output only if all inputs are logic "0".
   B. Produces a logic "1" at its output only if all inputs are logic "1".
   C. Produces a logic "0" at its output if any or all inputs are logic "1".
   D. Produces a logic "1" at its output if some but not all of its inputs are logic "1".

5. What is a characteristic of a NOT gate? **B**

   A. Does not allow data transmission when its input is high.
   B. Produces a logic "0" at its output when the input is logic "1" and vice versa.
   C. Allows data transmission only when its input is high.
   D. Produces a logic "1" at its output when the input is logic "1" and vice versa.

6. Which of the following logic gates will provide an active high out when both inputs are active high? **C**

   A. NAND.
   B. NOR.
   C. AND.
   D. XOR.

Key Topic 35: Logic Levels

1 In a negative-logic circuit, what level is used to represent a logic 0? **D**

A. Low level.
B. Positive-transition level.
C. Negative-transition level.
D. High level.

2. For the logic input levels shown in Figure 3E12, what are the logic levels of test points A, B and C in this circuit? (Assume positive logic.) **B**

A. A is high, B is low and C is low.
B. A is low, B is high and C is high.
C. A is high, B is high and C is low.
D. A is low, B is high and C is low.

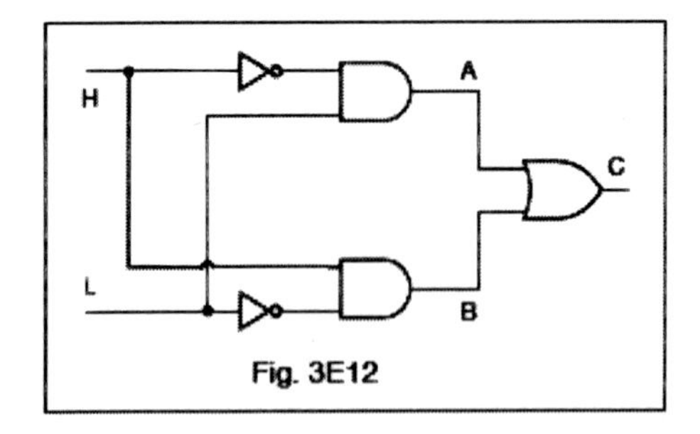

Fig. 3E12

3. For the logic input levels given in Figure 3E13, what are the logic levels of test points A, B and C in this circuit? (Assume positive logic.) C

A. A is low, B is low and C is high.
B. A is low, B is high and C is low.
C. A is high, B is high and C is high.
D. A is high, B is low and C is low.

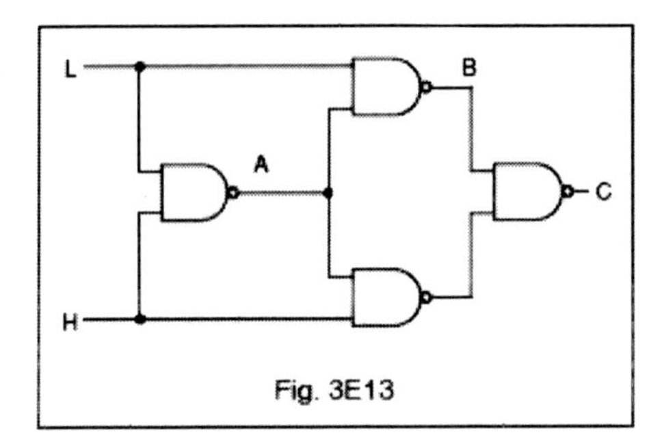

Fig. 3E13

4. In a positive-logic circuit, what level is used to represent a logic 1? **A**

A. High level
B. Low level
C. Positive-transition level
D. Negative-transition level

5. Given the input levels shown in Figure 3E14 and assuming positive logic devices, what would the output be? **A**

A. A is low, B is high and C is high.
B. A is high, B is high and C is low.
C. A is low, B is low and C is high.
D. None of the above are correct.

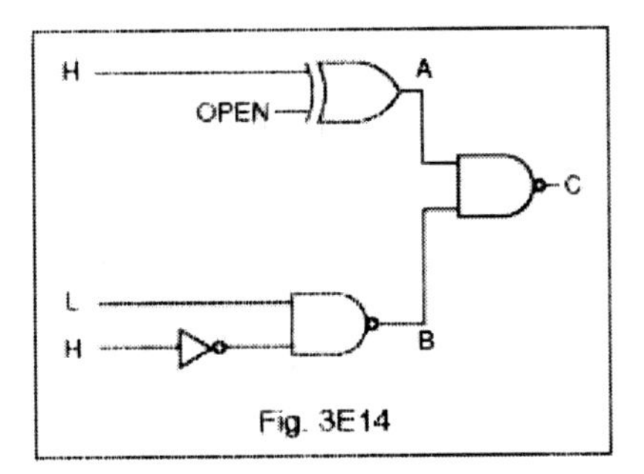

Fig. 3E14

6. What is a truth table? **A**

A. A list of input combinations and their corresponding outputs that characterizes a digital device's function.
B. A table of logic symbols that indicate the high logic states of an op-amp.
C. A diagram showing logic states when the digital device's output is true.
D. A table of logic symbols that indicates the low logic states of an op-amp.

Key Topic 36: Flip-Flops

1. A flip-flop circuit is a binary logic element with how many stable states? **B**
   - A. 1
   - B. 2
   - C. 4
   - D. 8

2. What is a flip-flop circuit? A binary sequential logic element with ___stable states. **C**
   - A. 1
   - B. 4
   - C. 2
   - D. 8

3. How many flip-flops are required to divide a signal frequency by 4? **D**
   - A. 1
   - B. 4
   - C. 8
   - D. 2

4. How many bits of information can be stored in a single flip-flop circuit? **A**
   - A. 1
   - B. 2
   - C. 3
   - D. 4

5. How many R-S flip-flops would be required to construct an 8 bit storage register? **C**
   - A. 2
   - B. 4
   - C. 8
   - D. 16

6. An R-S flip-flop is capable of doing all of the following except: **D**
   - A. Accept data input into R-S inputs with CLK initiated.
   - B. Accept data input into PRE and CLR inputs without CLK being initiated.
   - C. Refuse to accept synchronous data if asynchronous data is being input at same time.
   - D. Operate in toggle mode with R-S inputs held constant and CLK initiated.

Key Topic 37: Multivibrators

1. The frequency of an AC signal can be divided electronically by what type of digital circuit? **B**

   A. Free-running multivibrator.
   B. Bistable multivibrator.
   C. OR gate.
   D. Astable multivibrator.

2. What is an astable multivibrator? **D**

   A. A circuit that alternates between two stable states.
   B. A circuit that alternates between a stable state and an unstable state.
   C. A circuit set to block either a 0 pulse or a 1 pulse and pass the other.
   D. A circuit that alternates between two unstable states.

3. What is a monostable multivibrator? **A**

   A. A circuit that can be switched momentarily to the opposite binary state and then returns after a set time to its original state.
   B. A "clock" circuit that produces a continuous square wave oscillating between 1 and 0.
   C. A circuit designed to store one bit of data in either the 0 or the 1 configuration.
   D. A circuit that maintains a constant output voltage, regardless of variations in the input voltage.

4. What is a bistable multivibrator circuit commonly named? **D**

   A. AND gate.
   B. OR gate.
   C. Clock.
   D. Flip-flop.

5. What is a bistable multivibrator circuit? **A**

   A. Flip-flop.
   B. AND gate.
   C. OR gate.
   D. Clock.

6. What wave form would appear on the voltage outputs at the collectors of an astable, multivibrator, common-emitter stage? **C**

   A. Sine wave.
   B. Sawtooth wave.
   C. Square wave.
   D. Half-wave pulses.

Key Topic 38: Memory

1. What is the name of the semiconductor memory IC whose digital data can be written or read, and whose memory word address can be accessed randomly? **C**

   A. ROM – Read-Only Memory.
   B. PROM – Programmable Read-Only Memory.
   C. RAM – Random-Access Memory.
   D. EPROM – Electrically Programmable Read-Only Memory.

2. What is the name of the semiconductor IC that has a fixed pattern of digital data stored in its memory matrix? **B**

   A. RAM – Random-Access Memory.
   B. ROM – Read-Only Memory.
   C. Register.
   D. Latch.

3. What does the term "IO" mean within a microprocessor system? **C**

   A. Integrated oscillator.
   B. Integer operation.
   C. Input-output.
   D Internal operation.

4. What is the name for a microprocessor's sequence of commands and instructions? **A**

   A. Program.
   B. Sequence.
   C. Data string.
   D. Data execution.

5. How many individual memory cells would be contained in a memory IC that has 4 data bus input/output pins and 4 address pins for connection to the address bus? **D**

   A. 8
   B. 16
   C. 32
   D. 64

6 What is the name of the random-accessed semiconductor memory IC that must be refreshed periodically to maintain reliable data storage in its memory matrix? **B**

   A. ROM – Read-Only Memory.
   B. DRAM – Dynamic Random-Access Memory.
   C. PROM – Programmable Read-Only Memory.
   D. PRAM – Programmable Random-Access Memory.

Key Topic 39: Microprocessors

1. In a microprocessor-controlled two-way radio, a "watchdog" timer: **A**

   A. Verifies that the microprocessor is executing the program.
   B. Assures that the transmission is exactly on frequency.
   C. Prevents the transmitter from exceeding allowed power out.
   D. Connects to the system RADAR presentation.

2. What does the term "DAC" refer to in a microprocessor circuit? **B**

   A. Dynamic access controller.
   B. Digital to analog converter.
   C. Digital access counter.
   D. Dial analog control.

3. Which of the following is not part of a MCU processor? **D**

   A. RAM
   B. ROM
   C. I/O
   D. Voltage Regulator

4. What portion of a microprocessor circuit is the pulse generator? **A**

   A. Clock
   B. RAM
   C. ROM
   D. PLL

5. In a microprocessor, what is the meaning of the term "ALU"? **B**

   A. Automatic lock/unlock.
   B. Arithmetical logic unit.
   C. Auto latch undo.
   D. Answer local unit.

6. What circuit interconnects the microprocessor with the memory and input/output system? **C**

   A. Control logic bus.
   B. PLL line.
   C. Data bus line.
   D. Directional coupler.

Key Topic 40: Counters, Dividers, Converters

1. What is the purpose of a prescaler circuit? **D**
   - A. Converts the output of a JK flip-flop to that of an RS flip-flop.
   - B. Multiplies an HF signal so a low-frequency counter can display the operating frequency.
   - C. Prevents oscillation in a low frequency counter circuit.
   - D. Divides an HF signal so that a low-frequency counter can display the operating frequency.

2. What does the term "BCD" mean? **C**
   - A. Binaural coded digit.
   - B. Bit count decimal.
   - C. Binary coded decimal.
   - D. Broad course digit.

3. What is the function of a decade counter digital IC? **B**
   - A. Decode a decimal number for display on a seven-segment LED display.
   - B. Produce one output pulse for every ten input pulses.
   - C. Produce ten output pulses for every input pulse.
   - D. Add two decimal numbers.

4. What integrated circuit device converts an analog signal to a digital signal? **C**
   - A. DAC
   - B. DCC
   - C. ADC
   - D. CDC

5. What integrated circuit device converts digital signals to analog signals? **D**
   - A. ADC
   - B. DCC
   - C. CDC
   - D. DAC

6. In binary numbers, how would you note the quantity TWO? **A**
   - A. 0010
   - B. 0002
   - C. 2000
   - D. 0020

Subelement F – Receivers: 10 Key Topics - 10 Exam Questions

Key Topic 41: Receiver Theory

1. What is the limiting condition for sensitivity in a communications receiver? **A**

   A. The noise floor of the receiver.
   B. The power supply output ripple.
   C. The two-tone intermodulation distortion.
   D. The input impedance to the detector.

2. What is the definition of the term "receiver desensitizing"? **B**

   A. A burst of noise when the squelch is set too low.
   B. A reduction in receiver sensitivity because of a strong signal on a nearby frequency.
   C. A burst of noise when the squelch is set too high.
   D. A reduction in receiver sensitivity when the AF gain control is turned down.

3. What is the term used to refer to a reduction in receiver sensitivity caused by unwanted high-level adjacent channel signals? **A**

   A. Desensitizing.
   B. Intermodulation distortion.
   C. Quieting.
   D. Overloading.

4. What is meant by the term noise figure of a communications receiver? **C**

   A. The level of noise entering the receiver from the antenna.
   B. The relative strength of a received signal 3 kHz removed from the carrier frequency.
   C. The level of noise generated in the front end and succeeding stages of a receiver.
   D. The ability of a receiver to reject unwanted signals at frequencies close to the desired one.

5. Which stage of a receiver primarily establishes its noise figure? **B**

   A. The audio stage.
   B. The RF stage.
   C. The IF strip.
   D. The local oscillator.

6. What is the term for the ratio between the largest tolerable receiver input signal and the minimum discernible signal? **D**

   A. Intermodulation distortion.
   B. Noise floor.
   C. Noise figure.
   D. Dynamic range.

Key Topic 42: RF Amplifiers

1. How can selectivity be achieved in the front-end circuitry of a communications receiver? **D**
   A. By using an audio filter.
   B. By using an additional RF amplifier stage.
   C. By using an additional IF amplifier stage.
   D. By using a preselector.

2. What is the primary purpose of an RF amplifier in a receiver? **C**
   A. To provide most of the receiver gain.
   B. To vary the receiver image rejection by utilizing the AGC.
   C. To improve the receiver's noise figure.
   D. To develop the AGC voltage.

3. How much gain should be used in the RF amplifier stage of a receiver? **A**
   A. Sufficient gain to allow weak signals to overcome noise generated in the first mixer stage.
   B. As much gain as possible short of self oscillation.
   C. Sufficient gain to keep weak signals below the noise of the first mixer stage.
   D. It depends on the amplification factor of the first IF stage.

4. Too much gain in a VHF receiver front end could result in this: **D**
   A. Local signals become weaker.
   B. Difficult to match receiver impedances.
   C. Dramatic increase in receiver current.
   D. Susceptibility of intermodulation interference from nearby transmitters.

5. What is the advantage of a GaAsFET preamplifier in a modern VHF radio receiver? **C**
   A. Increased selectivity and flat gain.
   B. Low gain but high selectivity.
   C. High gain and low noise floor.
   D. High gain with high noise floor.

6. In what stage of a VHF receiver would a low noise amplifier be most advantageous? **B**
   A. IF stage.
   B. Front end RF stage.
   C. Audio stage.
   D. Power supply.

Key Topic 43: Oscillators

1. Why is the Colpitts oscillator circuit commonly used in a VFO (variable frequency oscillator)? **C**

   A. It can be phase locked.
   B. It can be remotely tuned.
   C. It is stable.
   D. It has little or no effect on the crystal's stability.

2. What is the oscillator stage called in a frequency synthesizer? **A**

   A. VCO.
   B. Divider.
   C. Phase detector.
   D. Reference standard.

3. What are three major oscillator circuits found in radio equipment? **D**

   A. Taft, Pierce, and negative feedback.
   B. Colpitts, Hartley, and Taft.
   C. Taft, Hartley, and Pierce.
   D. Colpitts, Hartley, and Pierce.

4. Which type of oscillator circuit is commonly used in a VFO (variable frequency oscillator)? **A**

   A. Colpitts.
   B. Pierce.
   C. Hartley.
   D. Negative feedback.

5. What condition must exist for a circuit to oscillate? It must: **D**

   A. Have a gain of less than 1.
   B. Be neutralized.
   C. Have sufficient negative feedback.
   D. Have sufficient positive feedback.

6. In Figure 3F15, which block diagram symbol (labeled 1 through 4) is used to represent a local oscillator? **B**

   A. 1
   B. 2
   C. 3
   D. 4

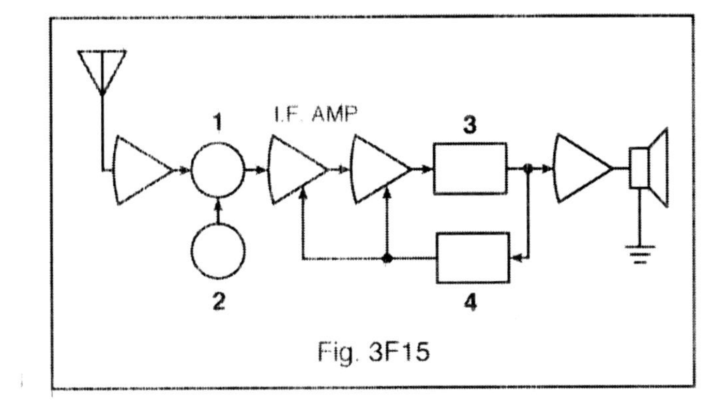

Fig. 3F15

Key Topic 44: Mixers

1. What is the image frequency if the normal channel is 151.000 MHz, the IF is operating at 11.000 MHz, and the LO is at 140.000 MHz? **B**

   A. 131.000 MHz.
   B. 129.000 MHz.
   C. 162.000 MHz.
   D. 150.000 MHz.

2. What is the mixing process in a radio receiver? **D**

   A. The elimination of noise in a wideband receiver by phase comparison.
   B. The elimination of noise in a wideband receiver by phase differentiation.
   C. Distortion caused by auroral propagation.
   D. The combination of two signals to produce sum and difference frequencies.

3. In what radio stage is the image frequency normally rejected? **A**

   A. RF.
   B. IF.
   C. LO.
   D. Detector.

4. What are the principal frequencies that appear at the output of a mixer circuit? **C**

   A. Two and four times the original frequency.
   B. The sum, difference and square root of the input frequencies.
   C. The original frequencies and the sum and difference frequencies.
   D. 1.414 and 0.707 times the input frequency.

5. If a receiver mixes a 13.8 MHz VFO with a 14.255 MHz receive signal to produce a 455 kHz intermediate frequency signal, what type of interference will a 13.345 MHz signal produce in the receiver? **B**

   A. Local oscillator interference.
   B. An image response.
   C. Mixer interference.
   D. Intermediate frequency interference.

6. What might occur in a receiver if excessive amounts of signal energy overdrive the mixer circuit? **C**

   A. Automatic limiting occurs.
   B. Mixer blanking occurs.
   C. Spurious mixer products are generated.
   D. The mixer circuit becomes unstable and drifts.

Key Topic 45: IF Amplifiers

1. What degree of selectivity is desirable in the IF circuitry of a wideband FM phone receiver? **D**
   - A. 1 kHz.
   - B. 2.4 kHz.
   - C. 4.2 kHz.
   - D. 15 kHz.

2. Which one of these filters can be used in micro-miniature electronic circuits? **B**
   - A. High power transmitter cavity.
   - B. Receiver SAW IF filter.
   - C. Floppy disk controller.
   - D. Internet DSL to telephone line filter.

3. A receiver selectivity of 2.4 kHz in the IF circuitry is optimum for what type of signals? **C**
   - A. CW.
   - B. Double-sideband AM voice.
   - C. SSB voice.
   - D. FSK RTTY.

4. A receiver selectivity of 10 KHz in the IF circuitry is optimum for what type of signals? **A**
   - A. Double-sideband AM.
   - B. SSB voice.
   - C. CW.
   - D. FSK RTTY.

5. What is an undesirable effect of using too wide a filter bandwidth in the IF section of a receiver? **B**
   - A. Output-offset overshoot.
   - B. Undesired signals will reach the audio stage.
   - C. Thermal-noise distortion.
   - D. Filter ringing.

6. How should the filter bandwidth of a receiver IF section compare with the bandwidth of a received signal? **A**
   - A. Slightly greater than the received-signal bandwidth.
   - B. Approximately half the received-signal bandwidth.
   - C. Approximately two times the received-signal bandwidth.
   - D. Approximately four times the received-signal bandwidth.

Key Topic 46: Filters and IF Amplifiers

1. What is the primary purpose of the final IF amplifier stage in a receiver? **B**
   - A. Dynamic response.
   - B. Gain.
   - C. Noise figure performance.
   - D. Bypass undesired signals.

2. What factors should be considered when selecting an intermediate frequency? **C**
   - A. Cross-modulation distortion and interference.
   - B. Interference to other services.
   - C. Image rejection and selectivity.
   - D. Noise figure and distortion.

3. What is the primary purpose of the first IF amplifier stage in a receiver? **D**
   - A. Noise figure performance.
   - B. Tune out cross-modulation distortion.
   - C. Dynamic response.
   - D. Selectivity.

4. What parameter must be selected when designing an audio filter using an op-amp? **A**
   - A. Bandpass characteristics.
   - B. Desired current gain.
   - C. Temperature coefficient.
   - D. Output-offset overshoot.

5. What are the distinguishing features of a Chebyshev filter? **C**
   - A. It has a maximally flat response over its passband.
   - B. It only requires inductors.
   - C. It allows ripple in the passband.
   - D. A filter whose product of the series- and shunt-element impedances is a constant for all frequencies.

6. When would it be more desirable to use an m-derived filter over a constant-k filter? **D**
   - A. When the response must be maximally flat at one frequency.
   - B. When the number of components must be minimized.
   - C. When high power levels must be filtered.
   - D. When you need more attenuation at a certain frequency that is too close to the cut-off frequency for a constant-k filter.

Key Topic 47: Filters

1. A good crystal band-pass filter for a single-sideband phone would be? **B**

A. 5 KHz.
B. 2.1 KHz.
C. 500 Hz.
D. 15 KHz.

2. Which statement is true regarding the filter output characteristics shown in Figure 3F16? **D**

A. C is a low pass curve and B is a band pass curve.
B. B is a high pass curve and D is a low pass curve.
C. A is a high pass curve and B is a low pass curve.
D. A is a low pass curve and D is a band stop curve.

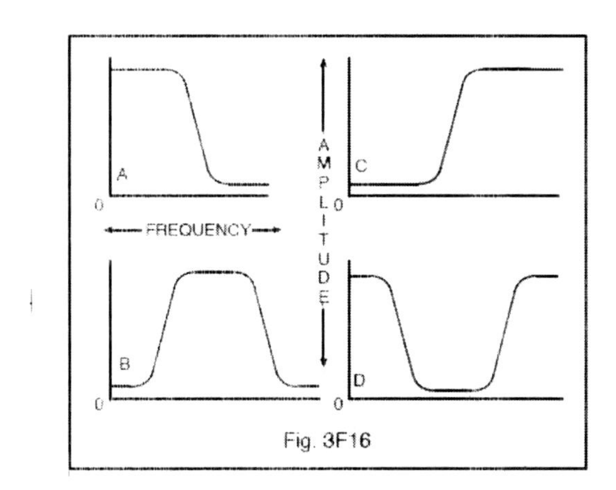

Fig. 3F16

3. What are the three general groupings of filters? **A**

A. High-pass, low-pass and band-pass.
B. Inductive, capacitive and resistive.
C. Audio, radio and capacitive.
D. Hartley, Colpitts and Pierce.

4. What is an m-derived filter? **D**

A. A filter whose input impedance varies widely over the design bandwidth.
B. A filter whose product of the series- and shunt-element impedances is a constant for all frequencies.
C. A filter whose schematic shape is the letter "M".
D. A filter that uses a trap to attenuate undesired frequencies too near cutoff for a constant-k filter.

5. What is an advantage of a constant-k filter? **A**

A. It has high attenuation of signals at frequencies far removed from the pass band.
B. It can match impedances over a wide range of frequencies.
C. It uses elliptic functions.
D. The ratio of the cutoff frequency to the trap frequency can be varied.

6. What are the distinguishing features of a Butterworth filter? **C**

A. A filter whose product of the series- and shunt-element impedances is a constant for all frequencies.
B. It only requires capacitors.
C. It has a maximally flat response over its passband.
D. It requires only inductors.

Key Topic 48: Detectors

1. What is a product detector? **C**
   A. It provides local oscillations for input to the mixer.
   B. It amplifies and narrows the band-pass frequencies.
   C. It uses a mixing process with a locally generated carrier.
   D. It is used to detect cross-modulation products.

2. Which circuit is used to detect FM-phone signals? **B**
   A. Balanced modulator.
   B. Frequency discriminator.
   C. Product detector.
   D. Phase splitter.

3. What is the process of detection in a radio diode detector circuit? **C**
   A. Breakdown of the Zener voltage.
   B. Mixing with noise in the transition region of the diode.
   C. Rectification and filtering of RF.
   D. The change of reactance in the diode with respect to frequency.

4. What is a frequency discriminator in a radio receiver? **A**
   A. A circuit for detecting FM signals.
   B. A circuit for filtering two closely adjacent signals.
   C. An automatic band switching circuit.
   D. An FM generator.

5. In a CTCSS controlled FM receiver, the CTCSS tone is filtered out after the: **D**
   A. IF stage but before the mixer.
   B. Mixer but before the IF.
   C. IF but before the discriminator.
   D. Discriminator but before the audio section.

6. What is the definition of detection in a radio receiver? **B**
   A. The process of masking out the intelligence on a received carrier to make an S-meter operational.
   B. The recovery of intelligence from the modulated RF signal.
   C. The modulation of a carrier.
   D. The mixing of noise with the received signal.

Key Topic 49: Audio & Squelch Circuits

1. What is the digital signal processing term for noise subtraction circuitry? **A**
   A. Adaptive filtering and autocorrelation.
   B. Noise blanking.
   C. Noise limiting.
   D. Auto squelch noise reduction.

2. What is the purpose of de-emphasis in the receiver audio stage? **B**
   A. When coupled with the transmitter pre-emphasis, flat audio is achieved.
   B. When coupled with the transmitter pre-emphasis, flat audio and noise reduction is received.
   C. No purpose is achieved.
   D. To conserve bandwidth by squelching no-audio periods in the transmission.

3. What makes a Digital Coded Squelch work? **D**
   A. Noise.
   B. Tones.
   C. Absence of noise.
   D. Digital codes.

4. What causes a squelch circuit to function? **A**
   A. Presence of noise.
   B. Absence of noise.
   C. Received tones.
   D. Received digital codes.

5. What makes a CTCSS squelch work? **B**
   A. Noise.
   B. Tones.
   C. Absence of noise.
   D. Digital codes.

6. What radio circuit samples analog signals, records and processes them as numbers, then converts them back to analog signals? **C**
   A. The pre-emphasis audio stage.
   B. The squelch gate circuit.
   C. The digital signal processing circuit.
   D. The voltage controlled oscillator circuit.

Key Topic 50: Receiver Performance

1. Where would you normally find a low-pass filter in a radio receiver? **D**

   A. In the AVC circuit.
   B. In the Oscillator stage.
   C. In the Power Supply.
   D. A and C, but not B.

2. How can ferrite beads be used to suppress ignition noise? Install them: **C**

   A. In the resistive high voltage cable every 2 years.
   B. Between the starter solenoid and the starter motor.
   C. Install them in the primary and secondary ignition leads.
   D. In the antenna lead.

3. What is the term used to refer to the condition where the signals from a very strong station are superimposed on other signals being received? **B**

   A. Intermodulation distortion.
   B. Cross-modulation interference.
   C. Receiver quieting.
   D. Capture effect.

4. What is cross-modulation interference? **C**

   A. Interference between two transmitters of different modulation type.
   B. Interference caused by audio rectification in the receiver preamp.
   C. Modulation from an unwanted signal heard in addition to the desired signal.
   D. Harmonic distortion of the transmitted signal.

5. In Figure 3F15 at what point in the circuit (labeled 1 through 4) could a DC voltmeter be used to monitor signal strength? **D**

   A. 1
   B. 2
   C. 3
   D. 4

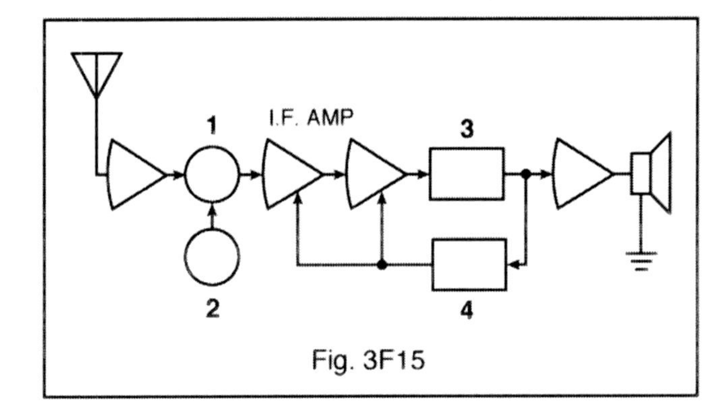

Fig. 3F15

6. Pulse type interference to automobile radio receivers that appears related to the speed of the engine can often be reduced by: **A**

   A. Installing resistances in series with spark plug wires.
   B. Using heavy conductors between the starting battery and the starting motor.
   C. Connecting resistances in series with the battery.
   D. Grounding the negative side of the battery.

Subelement G – Transmitters: 6 Key Topics - 6 Exam Questions

Key Topic 51: Amplifiers-1

1. What class of amplifier is distinguished by the presence of output throughout the entire signal cycle and the input never goes into the cutoff region? **A**
   - A. Class A.
   - B. Class B.
   - C. Class C.
   - D. Class D.

2. What is the distinguishing feature of a Class A amplifier? **B**
   - A. Output for less than 180 degrees of the signal cycle.
   - B. Output for the entire 360 degrees of the signal cycle.
   - C. Output for more than 180 degrees and less than 360 degrees of the signal cycle.
   - D. Output for exactly 180 degrees of the input signal cycle.

3. Which class of amplifier has the highest linearity and least distortion? **A**
   - A. Class A.
   - B. Class B.
   - C. Class C.
   - D. Class AB.

4. Which class of amplifier provides the highest efficiency? **C**
   - A. Class A.
   - B. Class B.
   - C. Class C.
   - D. Class AB.

5. What class of amplifier is distinguished by the bias being set well beyond cutoff? **B**
   - A. Class A.
   - B. Class C.
   - C. Class B.
   - D. Class AB.

6. Which class of amplifier has an operating angle of more than 180 degrees but less than 360 degrees when driven by a sine wave signal? **D**
   - A. Class A.
   - B. Class B.
   - C. Class C.
   - D. Class AB.

Key Topic 52: Amplifiers-2

1. The class B amplifier output is present for what portion of the input cycle? **D**

   A. 360 degrees.
   B. Greater than 180 degrees and less than 360 degrees.
   C. Less than 180 degrees.
   D. 180 degrees.

2. What input-amplitude parameter is most valuable in evaluating the signal-handling capability of a Class A amplifier? **C**

   A. Average voltage.
   B. RMS voltage.
   C. Peak voltage.
   D. Resting voltage.

3. The class C amplifier output is present for what portion of the input cycle? **A**

   A. Less than 180 degrees.
   B. Exactly 180 degrees.
   C. 360 degrees.
   D. More than 180 but less than 360 degrees.

4. What is the approximate DC input power to a Class AB RF power amplifier stage in an unmodulated carrier transmitter when the PEP output power is 500 watts? **D**

   A. 250 watts.
   B. 600 watts.
   C. 800 watts.
   D. 1000 watts.

5. The class AB amplifier output is present for what portion of the input cycle? **C**

   A. Exactly 180 degrees.
   B. 360 degrees
   C. More than 180 but less than 360 degrees.
   D. Less than 180 degrees.

6. What class of amplifier is characterized by conduction for 180 degrees of the input wave? **B**

   A. Class A.
   B. Class B.
   C. Class C.
   D. Class D.

Key Topic 53: Oscillators & Modulators

1. What is the modulation index in an FM phone signal having a maximum frequency deviation of 3,000 Hz on either side of the carrier frequency when the modulating frequency is 1,000 Hz? **C**
   - A. 0.3
   - B. 3,000
   - C. 3
   - D. 1,000

2. What is the modulation index of a FM phone transmitter producing a maximum carrier deviation of 6 kHz when modulated with a 2 kHz modulating frequency? **A**
   - A. 3
   - B. 6,000
   - C. 2,000
   - D. 1

3. What is the total bandwidth of a FM phone transmission having a 5 kHz deviation and a 3 kHz modulating frequency? **D**
   - A. 3 kHz.
   - B. 8 kHz.
   - C. 5 kHz.
   - D. 16 kHz.

4. How does the modulation index of a phase-modulated emission vary with RF carrier frequency? **A**
   - A. It does not depend on the RF carrier frequency.
   - B. Modulation index increases as the RF carrier frequency increases.
   - C. It varies with the square root of the RF carrier frequency.
   - D. It decreases as the RF carrier frequency increases.

5. How can a single-sideband phone signal be generated? **D**
   - A. By driving a product detector with a DSB signal.
   - B. By using a reactance modulator followed by a mixer.
   - C. By using a loop modulator followed by a mixer.
   - D. By using a balanced modulator followed by a filter.

6. What is a balanced modulator? **B**
   - A. An FM modulator that produces a balanced deviation.
   - B. A modulator that produces a double sideband, suppressed carrier signal.
   - C. A modulator that produces a single sideband, suppressed carrier signal.
   - D. A modulator that produces a full carrier signal.

Key Topic 54: Resonance - Tuning Networks

1. What is an L-network? **B**
   - A. A low power Wi-Fi RF network connection.
   - B. A network consisting of an inductor and a capacitor.
   - C. A "lossy" network.
   - D. A network formed by joining two low pass filters.

2. What is a pi-network? **D**
   - A. A network consisting of a capacitor, resistor and inductor.
   - B. The Phase inversion stage.
   - C. An enhanced token ring network.
   - D. A network consisting of one inductor and two capacitors or two inductors and onecapacitor.

3. What is the resonant frequency in an electrical circuit? **A**
   - A. The frequency at which capacitive reactance equals inductive reactance.
   - B. The highest frequency that will pass current.
   - C. The lowest frequency that will pass current.
   - D. The frequency at which power factor is at a minimum.

4. Which three network types are commonly used to match an amplifying device to a transmission line? **C**
   - A. Pi-C network, pi network and T network.
   - B. T network, M network and Z network.
   - C. L network, pi network and pi-L network.
   - D. L network, pi network and C network.

5. What is a pi-L network? **B**
   - A. A Phase Inverter Load network.
   - B. A network consisting of two inductors and two capacitors.
   - C. A network with only three discrete parts.
   - D. A matching network in which all components are isolated from ground.

6. Which network provides the greatest harmonic suppression? **C**
   - A. L network.
   - B. Pi network.
   - C. Pi-L network.
   - D. Inverse L network.

Key Topic 55: SSB Transmitters

1. What will occur when a non-linear amplifier is used with a single-sideband phone transmitter? **D**

   A. Reduced amplifier efficiency.
   B. Increased intelligibility.
   C. Sideband inversion.
   D. Distortion.

2. To produce a single-sideband suppressed carrier transmission it is necessary to ____ the carrier and to ____ the unwanted sideband. **B**

   A. Filter, filter.
   B. Cancel, filter.
   C. Filter, cancel.
   D. Cancel, cancel.

3. In a single-sideband phone signal, what determines the PEP-to-average power ratio? **C**

   A. The frequency of the modulating signal.
   B. The degree of carrier suppression.
   C. The speech characteristics.
   D. The amplifier power.

4. What is the approximate ratio of peak envelope power to average power during normal voice modulation peak in a single-sideband phone signal? **A**

   A. 2.5 to 1.
   B. 1 to 1.
   C. 25 to 1.
   D. 100 to 1.

5. What is the output peak envelope power from a transmitter as measured on an oscilloscope showing 200 volts peak-to-peak across a 50-ohm load resistor? **B**

   A. 1,000 watts.
   B. 100 watts.
   C. 200 watts.
   D. 400 watts.

6. What would be the voltage across a 50-ohm dummy load dissipating 1,200 watts? **A**

   A. 245 volts.
   B. 692 volts.
   C. 346 volts.
   D. 173 volts.

Key Topic 56: Technology

1. How can intermodulation interference between two transmitters in close proximity often be reduced or eliminated? **B**

   A. By using a Class C final amplifier with high driving power.
   B. By installing a terminated circulator or ferrite isolator in the feed line to the transmitter and duplexer.
   C. By installing a band-pass filter in the antenna feed line.
   D. By installing a low-pass filter in the antenna feed line.

2. How can parasitic oscillations be eliminated in a power amplifier? **C**

   A. By tuning for maximum SWR.
   B. By tuning for maximum power output.
   C. By neutralization.
   D. By tuning the output.

3. What is the name of the condition that occurs when the signals of two transmitters in close proximity mix together in one or both of their final amplifiers, and unwanted signals at the sum and difference frequencies of the original transmissions are generated? **D**

   A. Amplifier desensitization.
   B. Neutralization.
   C. Adjacent channel interference.
   D. Intermodulation interference.

4. What term describes a wide-bandwidth communications system in which the RF carrier varies according to some pre-determined sequence? **A**

   A. Spread-spectrum communication.
   B. AMTOR.
   C. SITOR.
   D. Time-domain frequency modulation.

5. How can even-order harmonics be reduced or prevented in transmitter amplifier design? **C**

   A. By using a push-push amplifier.
   B. By operating class C.
   C. By using a push-pull amplifier.
   D. By operating class AB.

6. What is the modulation type that can be a frequency hopping of one carrier or multiple simultaneous carriers? **D**

   A. SSB.
   B. FM.
   C. OFSK.
   D. Spread spectrum.

Subelement H – Modulation: 3 Key Topics - 3 Exam Questions

Key Topic 57: Frequency Modulation

1. The deviation ratio is the: **B**
   - A. Audio modulating frequency to the center carrier frequency.
   - B. Maximum carrier frequency deviation to the highest audio modulating frequency.
   - C. Carrier center frequency to the audio modulating frequency.
   - D. Highest audio modulating frequency to the average audio modulating frequency.

2. What is the deviation ratio for an FM phone signal having a maximum frequency deviation of plus or minus 5 kHz and accepting a maximum modulation rate of 3 kHz? **D**
   - A. 60
   - B. 0.16
   - C. 0.6
   - D. 1.66

3. What is the deviation ratio of an FM-phone signal having a maximum frequency swing of plus or minus 7.5 kHz and accepting a maximum modulation rate of 3.5 kHz? **A**
   - A. 2.14
   - B. 0.214
   - C. 0.47
   - D. 47

4. How can an FM-phone signal be produced in a transmitter? **D**
   - A. By modulating the supply voltage to a class-B amplifier.
   - B. By modulating the supply voltage to a class-C amplifier.
   - C. By using a balanced modulator.
   - D. By feeding the audio directly to the oscillator.

5. What is meant by the term modulation index? **A**
   - A. The ratio between the deviation of a frequency modulated signal and the modulating frequency.
   - B. The processor index.
   - C. The FM signal-to-noise ratio.
   - D. The ratio of the maximum carrier frequency deviation to the highest audio modulating frequency.

6. In an FM-phone signal, what is the term for the maximum deviation from the carrier frequency divided by the maximum audio modulating frequency? **C**
   - A. Deviation index.
   - B. Modulation index.
   - C. Deviation ratio.
   - D. Modulation ratio.

Key Topic 58: SSB Modulation

1. In Figure 3H17, the block labeled 4 would indicate that this schematic is most likely a/an: **C**

   A. Audio amplifier.
   B. Shipboard RADAR.
   C. SSB radio transmitter.
   D. Wireless LAN (local area network) computer.

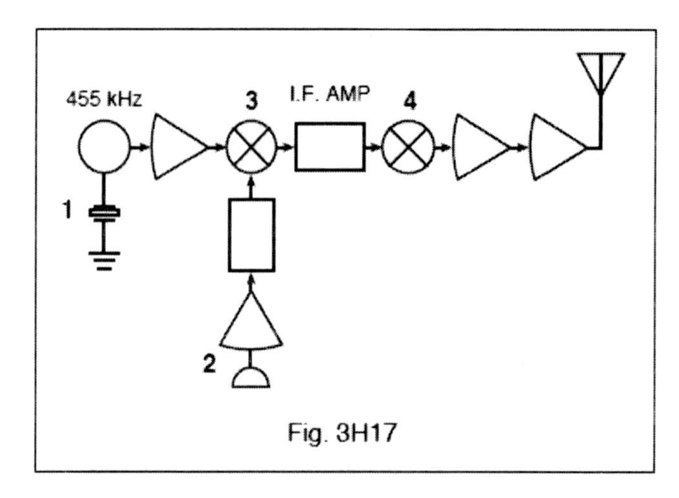

Fig. 3H17

2. In Figure 3H17, which block diagram symbol (labeled 1 through 4) represents where audio intelligence is inserted? **B**

   A. 1
   B. 2
   C. 3
   D. 4

3. What kind of input signal could be used to test the amplitude linearity of a single-sideband phone transmitter while viewing the output on an oscilloscope? **C**

   A. Whistling in the microphone.
   B. An audio frequency sine wave.
   C. A two-tone audio-frequency sine wave.
   D. An audio frequency square wave.

4. What does a two-tone test illustrate on an oscilloscope? **A**

   A. Linearity of a SSB transmitter.
   B. Frequency of the carrier phase shift.
   C. Percentage of frequency modulation.
   D. Sideband suppression.

5. How can a double-sideband phone signal be produced? **D**

   A. By using a reactance modulator.
   B. By varying the voltage to the varactor in an oscillator circuit.
   C. By using a phase detector, oscillator, and filter in a feedback loop.
   D. By modulating the supply voltage to a class C amplifier.

6. What type of signals are used to conduct an SSB two-tone test? **B**

   A. Two audio signals of the same frequency, but shifted 90 degrees in phase.
   B. Two non-harmonically related audio signals that are within the modulation band pass of the transmitter.
   C. Two different audio frequency square wave signals of equal amplitude.
   D. Any two audio frequencies as long as they are harmonically related.

Key Topic 59: Pulse Modulation

1. What is an important factor in pulse-code modulation using time-division multiplex? **A**
   - A. Synchronization of transmit and receive clock pulse rates.
   - B. Frequency separation.
   - C. Overmodulation and undermodulation.
   - D. Slight variations in power supply voltage.

2. In a pulse-width modulation system, what parameter does the modulating signal vary? **B**
   - A. Pulse frequency.
   - B. Pulse duration.
   - C. Pulse amplitude.
   - D. Pulse intensity.

3. What is the name of the type of modulation in which the modulating signal varies the duration of the transmitted pulse? **D**
   - A. Amplitude modulation.
   - B. Frequency modulation.
   - C. Pulse-height modulation.
   - D. Pulse-width modulation.

4. Which of the following best describes a pulse modulation system? **A**
   - A. The peak transmitter power is normally much greater than the average power.
   - B. Pulse modulation is sometimes used in SSB voice transmitters.
   - C. The average power is normally only slightly below the peak power.
   - D. The peak power is normally twice as high as the average power.

5. In a pulse-position modulation system, what parameter does the modulating signal vary? **B**
   - A. The number of pulses per second.
   - B. The time at which each pulse occurs.
   - C. Both the frequency and amplitude of the pulses.
   - D. The duration of the pulses.

6. What is one way that voice is transmitted in a pulse-width modulation system? **C**
   - A. A standard pulse is varied in amplitude by an amount depending on the voice waveform at that instant.
   - B. The position of a standard pulse is varied by an amount depending on the voice waveform at that instant.
   - C. A standard pulse is varied in duration by an amount depending on the voice waveform at that instant.
   - D. The number of standard pulses per second varies depending on the voice waveform at that instant.

Subelement I – Power Sources: 3 Key Topics - 3 Exam Questions

Key Topic 60: Batteries-1

1. When a lead-acid storage battery is being charged, a harmful effect to humans is: **D**
   A. Internal plate sulfation may occur under constant charging.
   B. Emission of oxygen.
   C. Emission of chlorine gas.
   D. Emission of hydrogen gas.

2. A battery with a terminal voltage of 12.5 volts is to be trickle-charged at a 0.5 A rate. What resistance should be connected in series with the battery to charge it from a 110-V DC line? **C**
   A. 95 ohms.
   B. 300 ohms.
   C. 195 ohms.
   D. None of these.

3. What capacity in amperes does a storage battery need to be in order to operate a 50 watt transmitter for 6 hours? Assume a continuous transmitter load of 70% of the key-locked demand of 40 A, and an emergency light load of 1.5 A. **B**
   A. 100 ampere-hours.
   B. 177 ampere-hours.
   C. 249 ampere-hours.
   D. None of these.

4. What is the total voltage when 12 Nickel-Cadmium batteries are connected in series? **C**
   A. 12 volts.
   B. 12.6 volts.
   C. 15 volts.
   D. 72 volts.

5. The average fully-charged voltage of a lead-acid storage cell is: **D**
   A. 1 volt.
   B. 1.2 volts.
   C. 1.56 volts.
   D. 2.06 volts.

6. A nickel-cadmium cell has an operating voltage of about: **A**
   A. 1.25 volts.
   B. 1.4 volts.
   C. 1.5 volts.
   D. 2.1 volts.

Key Topic 61: Batteries-2

1. When an emergency transmitter uses 325 watts and a receiver uses 50 watts, how many hours can a 12.6 volt, 55 ampere-hour battery supply full power to both units? **A**

   A. 1.8 hours.
   B. 6 hours.
   C. 3 hours.
   D. 1.2 hours.

2. What current will flow in a 6 volt storage battery with an internal resistance of 0.01 ohms, when a 3-watt, 6-volt lamp is connected? **B**

   A. 0.4885 amps.
   B. 0.4995 amps.
   C. 0.5566 amps.
   D. 0.5795 amps.

3. A ship RADAR unit uses 315 watts and a radio uses 50 watts. If the equipment is connected to a 50 ampere-hour battery rated at 12.6 volts, how long will the battery last? **A**

   A. 1 hour 43 minutes.
   B. 28.97 hours.
   C. 29 minutes.
   D. 10 hours, 50 minutes.

4. If a marine radiotelephone receiver uses 75 watts of power and a transmitter uses 325 watts, how long can they both operate before discharging a 50 ampere-hour 12 volt battery? **C**

   A. 40 minutes.
   B. 1 hour.
   C. 1 1/2 hours.
   D. 6 hours.

5. A 6 volt battery with 1.2 ohms internal resistance is connected across two light bulbs in parallel whose resistance is 12 ohms each. What is the current flow? **B**

   A. 0.57 amps.
   B. 0.83 amps.
   C. 1.0 amps.
   D. 6.0 amps.

6. A 12.6 volt, 8 ampere-hour battery is supplying power to a receiver that uses 50 watts and a RADAR system that uses 300 watts. How long will the battery last? **D**

   A. 100.8 hours.
   B. 27.7 hours.
   C. 1 hour.
   D. 17 minutes or 0.3 hours.

Key Topic 62: Motors & Generators

1. What occurs if the load is removed from an operating series DC motor? **D**
   A. It will stop running.
   B. Speed will increase slightly.
   C. No change occurs.
   D. It will accelerate until it falls apart.

2. If a shunt motor running with a load has its shunt field opened, how would this affect the speed of the motor? **C**
   A. It will slow down.
   B. It will stop suddenly.
   C. It will speed up.
   D. It will be unaffected.

3. The expression "voltage regulation" as it applies to a shunt-wound DC generator operating at a constant frequency refers to: **A**
   A. Voltage fluctuations from load to no-load.
   B. Voltage output efficiency.
   C. Voltage in the secondary compared to the primary.
   D. Rotor winding voltage ratio

4. What is the line current of a 7 horsepower motor operating on 120 volts at full load, a power factor of 0.8, and 95% efficient? **D**
   A. 4.72 amps.
   B. 13.03 amps.
   C. 56 amps.
   D. 57.2 amps.

5. A 3 horsepower, 100 V DC motor is 85% efficient when developing its rated output. What is the current? **C**
   A. 8.545 amps.
   B. 20.345 amps.
   C. 26.300 amps.
   D. 25.000 amps.

6. The output of a separately-excited AC generator running at a constant speed can be controlled by: **B**
   A. The armature.
   B. The amount of field current.
   C. The brushes.
   D. The exciter.

Subelement J -- Antennas: 5 Key Topics - 5 Exam Questions

Key Topic 63: Antenna Theory

1. Which of the following could cause a high standing wave ratio on a transmission line? **C**
   - A. Excessive modulation.
   - B. An increase in output power.
   - C. A detuned antenna coupler.
   - D. Low power from the transmitter.

2. Why is the value of the radiation resistance of an antenna important? **A**
   - A. Knowing the radiation resistance makes it possible to match impedances for maximum power transfer.
   - B. Knowing the radiation resistance makes it possible to measure the near-field radiation density from transmitting antenna.
   - C. The value of the radiation resistance represents the front-to-side ratio of the antenna.
   - D. The value of the radiation resistance represents the front-to-back ratio of the antenna.

3. A radio frequency device that allows RF energy to pass through in one direction with very little loss but absorbs RF power in the opposite direction is a: **D**
   - A. Circulator.
   - B. Wave trap.
   - C. Multiplexer.
   - D. Isolator.

4. What is an advantage of using a trap antenna? **A**
   - A. It may be used for multiband operation.
   - B. It has high directivity in the high-frequency bands.
   - C. It has high gain.
   - D. It minimizes harmonic radiation.

5. What is meant by the term radiation resistance of an antenna? **D**
   - A. Losses in the antenna elements and feed line.
   - B. The specific impedance of the antenna.
   - C. The resistance in the trap coils to received signals.
   - D. An equivalent resistance that would dissipate the same amount of power as that radiated from an antenna.

6. What is meant by the term antenna bandwidth? **B**
   - A. Antenna length divided by the number of elements.
   - B. The frequency range over which an antenna can be expected to perform well.
   - C. The angle between the half-power radiation points.
   - D. The angle formed between two imaginary lines drawn through the ends of the elements.

Key Topic 64: Voltage, Current and Power Relationships

1. What is the current flowing through a 52 ohm line with an input of 1,872 watts? **B**
   - A. 0.06 amps.
   - B. 6 amps.
   - C. 28.7 amps.
   - D. 144 amps.

2. The voltage produced in a receiving antenna is: **D**
   - A. Out of phase with the current if connected properly.
   - B. Out of phase with the current if cut to 1/3 wavelength.
   - C. Variable depending on the station's SWR.
   - D. Always proportional to the received field strength.

3. Which of the following represents the best standing wave ratio (SWR)? **A**
   - A. 1:1.
   - B. 1:1.5.
   - C. 1:3.
   - D. 1:4.

4. At the ends of a half-wave antenna, what values of current and voltage exist compared to the remainder of the antenna? **C**
   - A. Equal voltage and current.
   - B. Minimum voltage and maximum current.
   - C. Maximum voltage and minimum current.
   - D. Minimum voltage and minimum current.

5. An antenna radiates a primary signal of 500 watts output. If there is a 2nd harmonic output of 0.5 watt, what attenuation of the 2nd harmonic has occurred? **B**
   - A. 10 dB.
   - B. 30 dB.
   - C. 40 dB.
   - D. 50 dB.

6. There is an improper impedance match between a 30 watt transmitter and the antenna, with 5 watts reflected. How much power is actually radiated? **C**
   - A. 35 watts.
   - B. 30 watts.
   - C. 25 watts.
   - D. 20 watts.

Key Topic 65: Frequency and Bandwidth

1. A vertical 1/4 wave antenna receives signals: **D**

   A. In the microwave band.
   B. In one vertical direction.
   C. In one horizontal direction.
   D. Equally from all horizontal directions.

2. The resonant frequency of a Hertz antenna can be lowered by: **B**

   A. Lowering the frequency of the transmitter.
   B. Placing an inductance in series with the antenna.
   C. Placing a condenser in series with the antenna.
   D. Placing a resistor in series with the antenna.

3. An excited 1/2 wavelength antenna produces: **C**

   A. Residual fields.
   B. An electro-magnetic field only.
   C. Both electro-magnetic and electro-static fields.
   D. An electro-flux field sometimes.

4. To increase the resonant frequency of a 1/4 wavelength antenna: **A**

   A. Add a capacitor in series.
   B. Lower capacitor value.
   C. Cut antenna.
   D. Add an inductor.

5. What happens to the bandwidth of an antenna as it is shortened through the use of loading coils? **B**

   A. It is increased.
   B. It is decreased.
   C. No change occurs.
   D. It becomes flat.

6. To lengthen an antenna electrically, add a: **A**

   A. Coil.
   B. Resistor.
   C. Battery.
   D. Conduit.

Key Topic 66: Transmission Lines

1. What is the meaning of the term velocity factor of a transmission line? **B**
   - A. The ratio of the characteristic impedance of the line to the terminating impedance.
   - B. The velocity of the wave on the transmission line divided by the velocity of light in a vacuum.
   - C. The velocity of the wave on the transmission line multiplied by the velocity of light in a vacuum.
   - D. The index of shielding for coaxial cable.

2. What determines the velocity factor in a transmission line? **C**
   - A. The termination impedance.
   - B. The line length.
   - C. Dielectrics in the line.
   - D. The center conductor resistivity.

3. Nitrogen is placed in transmission lines to: **D**
   - A. Improve the "skin-effect" of microwaves.
   - B. Reduce arcing in the line.
   - C. Reduce the standing wave ratio of the line.
   - D. Prevent moisture from entering the line.

4. A perfect (no loss) coaxial cable has 7 dB of reflected power when the input is 5 watts. What is the output of the transmission line? **A**
   - A. 1 watt.
   - B. 1.25 watts.
   - C. 2.5 watts.
   - D. 5 watts.

5. Referred to the fundamental frequency, a shorted stub line attached to the transmission line to absorb even harmonics could have a wavelength of: **C**
   - A. 1.41 wavelength.
   - B. 1/2 wavelength.
   - C. 1/4 wavelength.
   - D. 1/6 wavelength.

6. If a transmission line has a power loss of 6 dB per 100 feet, what is the power at the feed point to the antenna at the end of a 200 foot transmission line fed by a 100 watt transmitter? **D**
   - A. 70 watts.
   - B. 50 watts.
   - C. 25 watts.
   - D. 6 watts.

Key Topic 67: Effective Radiated Power

1. What is the effective radiated power of a repeater with 50 watts transmitter power output, 4 dB feedline loss, 3 dB duplexer and circulator loss, and 6 dB antenna gain? **B**

   A. 158 watts.
   B. 39.7 watts.
   C. 251 watts.
   D. 69.9 watts.

2. What is the effective radiated power of a repeater with 75 watts transmitter power output, 4 dB feedline loss, 3 dB duplexer and circulator loss, and 10 dB antenna gain? **D**

   A. 600 watts.
   B. 75 watts.
   C. 18.75 watts.
   D. 150 watts.

3. What is the effective radiated power of a repeater with 75 watts transmitter power output, 5 dB feedline loss, 4 dB duplexer and circulator loss, and 6 dB antenna gain? **A**

   A. 37.6 watts.
   B. 237 watts.
   C. 150 watts.
   D. 23.7 watts.

4. What is the effective radiated power of a repeater with 100 watts transmitter power output, 4 dB feedline loss, 3 dB duplexer and circulator loss, and 7 dB antenna gain? **D**

   A. 631 watts.
   B. 400 watts.
   C. 25 watts.
   D. 100 watts.

5. What is the effective radiated power of a repeater with 100 watts transmitter power output, 5 dB feedline loss, 4 dB duplexer and circulator loss, and 10 dB antenna gain? **A**

   A. 126 watts.
   B. 800 watts.
   C. 12.5 watts.
   D. 1260 watts.

6. What is the effective radiated power of a repeater with 50 watts transmitter power output, 5 dB feedline loss, 4 dB duplexer and circulator loss, and 7 dB antenna gain? **C**

   A. 300 watts.
   B. 315 watts.
   C. 31.5 watts.
   D. 69.9 watts.

Subelement 3-K – Aircraft: 6 Key Topics - 6 Exam Questions

Key Topic 68: Distance Measuring Equipment

1. What is the frequency range of the Distance Measuring Equipment (DME) used to indicate an aircraft's slant range distance to a selected ground-based navigation station? **C**
   - A. 108.00 MHz to 117.95 MHz.
   - B. 108.10 MHz to 111.95 MHz.
   - C. 962 MHz to 1213 MHz.
   - D. 329.15 MHz to 335.00 MHz.

2. The Distance Measuring Equipment (DME) measures the distance from the aircraft to the DME ground station. This is referred to as: **B**
   - A. DME bearing.
   - B. The slant range.
   - C. Glide Slope angle of approach.
   - D. Localizer course width.

3. The Distance Measuring Equipment (DME) ground station has a built-in delay between reception of an interrogation and transmission of the reply to allow: **C**
   - A. Someone to answer the call.
   - B. The VOR to make a mechanical hook-up.
   - C. Operation at close range.
   - D. Clear other traffic for a reply.

4. What is the main underlying operating principle of an aircraft's Distance Measuring Equipment (DME)? **A**
   - A. A measurable amount of time is required to send and receive a radio signal hrough the Earth's atmosphere.
   - B. The difference between the peak values of two DC voltages may be used to determine an aircraft's distance to another aircraft.
   - C. A measurable frequency compression of an AC signal may be used to determine an aircraft's altitude above the earth.
   - D. A phase inversion between two AC voltages may be used to determine an aircraft's distance to the exit ramp of an airport's runway.

5. What radio navigation aid determines the distance from an aircraft to a selected VORTAC station by measuring the length of time the radio signal takes to travel to and from the station? **D**
   - A. RADAR.
   - B. Loran C.
   - C. Distance Marking (DM).
   - D. Distance Measuring Equipment (DME).

6. The majority of airborne Distance Measuring Equipment systems automatically tune their transmitter and receiver frequencies to the paired __ / __ channel. **B**
   - A. VOR/marker beacon.
   - B. VOR/LOC.
   - C. Marker beacon/glideslope.
   - D. LOC/glideslope.

Key Topic 69: VHF Omnidirectional Range (VOR)

1. All directions associated with a VOR station are related to: **A**

   A. Magnetic north.
   B. North pole.
   C. North star.
   D. None of these.

2. The rate that the transmitted VOR variable signal rotates is equivalent to how many revolutions per second? **B**

   A. 60
   B. 30
   C. 2400
   D. 1800

3. What is the frequency range of the ground-based Very-high-frequency Omnidirectional Range (VOR) stations used for aircraft navigation? **D**

   A. 108.00 kHz to 117.95 kHz.
   B. 329.15 MHz to 335.00 MHz.
   C. 329.15 kHz to 335.00 kHz.
   D. 108.00 MHz to 117.95 MHz.

4. Lines drawn from the VOR station in a particular magnetic direction are: **A**

   A. Radials.
   B. Quadrants.
   C. Bearings.
   D. Headings.

5. The amplitude modulated variable phase signal and the frequency modulated reference phase signal of a Very-high-frequency Omnidirectional Range (VOR) station used for aircraft navigation are synchronized so that both signals are in phase with each other at ___________ of the VOR station. **B**

   A. 180 degrees South, true bearing position.
   B. 360 degrees North, magnetic bearing position.
   C. 180 degrees South, magnetic bearing position.
   D. 0 degrees North, true bearing position.

6. What is the main underlying operating principle of the Very-high-frequency Omnidirectional Range (VOR) aircraft navigational system? **C**

   A. A definite amount of time is required to send and receive a radio signal.
   B. The difference between the peak values of two DC voltages may be used to a determine n aircraft's altitude above a selected VOR station.
   C. A phase difference between two AC voltages may be used to determine an aircraft's azimuth position in relation to a selected VOR station.
   D. A phase difference between two AC voltages may be used to determine an aircraft's distance from a selected VOR station.

Key Topic 70: Instrument Landing System (ILS)

1. What is the frequency range of the localizer beam system used by aircraft to find the centerline of a runway during an Instrument Landing System (ILS) approach to an airport? **D**

   A. 108.10 kHz to 111.95 kHz.
   B. 329.15 MHz to 335.00 MHz.
   C. 329.15 kHz to 335.00 kHz.
   D. 108.10 MHz to 111.95 MHz.

2. What is the frequency range of the marker beacon system used to indicate an aircraft's position during an Instrument Landing System (ILS) approach to an airport's runway? **C**

   A. The outer, middle, and inner marker beacons' UHF frequencies are unique for each ILS equipped airport to provide unambiguous frequency-protected reception areas in the 329.15 to 335.00 MHz range.
   B. The outer marker beacon's carrier frequency is 400 MHz, the middle marker beacon's carrier frequency is 1300 MHz, and the inner marker beacon's carrier frequency is 3000 MHz.
   C. The outer, the middle, and the inner marker beacon's carrier frequencies are all 75 MHz but the marker beacons are 95% tone-modulated at 400 Hz (outer), 1300 Hz (middle), and 3000 Hz (inner).
   D. The outer, marker beacon's carrier frequency is 3000 kHz, the middle marker beacon's carrier frequency is 1300 kHz, and the inner marker beacon's carrier frequency is 400 kHz.

3. Which of the following is a required component of an Instrument Landing System (ILS)? **B**

   A. Altimeter: shows aircraft height above sea-level.
   B. Localizer: shows aircraft deviation horizontally from center of runway.
   C. VHF Communications: provide communications to aircraft.
   D. Distance Measuring Equipment: shows aircraft distance to VORTAC station.

4. What type of antenna is used in an aircraft's Instrument Landing System (ILS) glideslope installation? **C**

   A. A vertically polarized antenna that radiates an omnidirectional antenna pattern.
   B. A balanced loop reception antenna.
   C. A folded dipole reception antenna.
   D. An electronically steerable phased-array antenna that radiates a directional antenna pattern.

5. Choose the only correct statement about the localizer beam system used by aircraft to find the centerline of a runway during an Instrument Landing System (ILS) approach to an airport. The localizer beam system: **D**

   A. Operates within the assigned frequency range of 108.10 to 111.95 GHz.
   B. Produces two amplitude modulated antenna patterns; one pattern above and one pattern below the normal 2.5 degree approach glide path of the aircraft.
   C. Frequencies are automatically tuned-in when the proper glide slope frequency is selected on the aircraft's Navigation and Communication (NAV/COMM) transceiver.
   D. Produces two amplitude modulated antenna patterns; one pattern with an audio frequency of 90 Hz and one pattern with an audio frequency of 150 Hz, one left of the runway centerline and one right of the runway centerline.

6. On runway approach, an ILS Localizer shows: **A**

   A. Deviation left or right of runway center line.
   B. Deviation up and down from ground speed.
   C. Deviation percentage from authorized ground speed.
   D. Wind speed along runway.

Key Topic 71: Automatic Direction Finding Equipment (ADF) & Transponders

1. What is the frequency range of an aircraft's Automatic Direction Finding (ADF) equipment? **A**
   A. 190 kHz to 1750 kHz.
   B. 190 MHz to 1750 MHz.
   C. 108.10 MHz to 111.95 MHz.
   D. 108.00 MHz to 117.95 MHz.

2. What is meant by the term "night effect" when using an aircraft's Automatic Direction Finding (ADF) equipment? Night effect refers to the fact that: **B**
   A. All Non Directional Beacon (NDB) transmitters are turned-off at dusk and turned-on at dawn.
   B. Non Directional Beacon (NDB) transmissions can bounce-off the Earth's ionosphere at night and be received at almost any direction.
   C. An aircraft's ADF transmissions will be slowed at night due to the increased density of the Earth's atmosphere after sunset.
   D. An aircraft's ADF antennas usually collect dew moisture after sunset which decreases their effective reception distance from an NDB transmitter.

3. What are the transmit and receive frequencies of an aircraft's mode C transponder operating in the Air Traffic Control RADAR Beacon System (ATCRBS)? **A**
   A. Transmit at 1090 MHz, and receive at 1030 MHz
   B. Transmit at 1030 kHz, and receive at 1090 kHz
   C. Transmit at 1090 kHz, and receive at 1030 kHz
   D. Transmit at 1030 MHz, and receive at 1090 MHz

4. In addition to duplicating the functions of a mode C transponder, an aircraft's mode S transponder can also provide: **C**
   A. Primary RADAR surveillance capabilities.
   B. Long range lightning detection.
   C. Mid-Air collision avoidance capabilities.
   D. Backup VHF voice communication abilities.

5. What type of encoding is used in an aircraft's mode C transponder transmission to a ground station of the Air Traffic Control RADAR Beacon System (ATCRBS)? **B**
   A. Differential phase shift keying.
   B. Pulse position modulation.
   C. Doppler effect compressional encryption.
   D. Amplitude modulation at 95%.

6. Choose the only correct statement about an aircraft's Automatic Direction Finding (ADF) equipment. **D**
   A. An aircraft's ADF transmission exhibits primarily a line-of-sight range to the ground-based target station and will not follow the curvature of the Earth.
   B. Only a single omnidirectional sense antenna is required to receive an NDB transmission and process the signal to calculate the aircraft's bearing to the selected ground station.
   C. All frequencies in the ADF's operating range except the commercial standard broadcast stations (550 to 1660 kHz) can be utilized as a navigational Non Directional Beacon (NDB) signal.
   D. An aircraft's ADF antennas can receive transmissions that are over the Earth's horizon (sometimes several hundred miles away) since these signals will follow the curvature of the Earth.

Key Topic 72: Aircraft Antenna Systems and Frequencies

1. What type of antenna pattern is radiated from a ground station phased-array directional antenna when transmitting the PPM pulses in a Mode S interrogation signal of an aircraft's Traffic alert and Collision Avoidance System (TCAS) installation? **D**
   A. 1090 MHz directional pattern.
   B. 1030 MHz omnidirectional pattern.
   C. 1090 MHz omnidirectional pattern.
   D. 1030 MHz directional pattern.

2. What type of antenna is used in an aircraft's Instrument Landing System (ILS) marker beacon installation? **C**
   A. An electronically steerable phased-array antenna that radiates adirectional antenna pattern.
   B. A folded dipole reception antenna.
   C. A balanced loop reception antenna.
   D. A horizontally polarized antenna that radiates an omnidirectional antenna pattern.

3. What is the frequency range of an aircraft's Very High Frequency (VHF) communications? **A**
   A. 118.000 MHz to 136.975 MHz (worldwide up to 151.975 MHz).
   B. 108.00 MHz to 117.95 MHz.
   C. 329.15 MHz to 335.00 MHz.
   D. 2.000 MHz to 29.999 MHz.

4. Aircraft Emergency Locator Transmitters (ELT) operate on what frequencies? **D**
   A. 121.5 MHz.
   B. 243 MHz.
   C. 121.5 and 243 MHz.
   D. 121.5, 243 and 406 MHz.

5. What is the frequency range of an aircraft's radio altimeter? **C**
   A. 962 MHz to 1213 MHz.
   B. 329.15 MHz to 335.00 MHz.
   C. 4250 MHz to 4350 MHz.
   D. 108.00 MHz to 117.95 MHz.

6. What type of antenna is attached to an aircraft's Mode C transponder installation and used to receive 1030 MHz interrogation signals from the Air Traffic Control Radar Beacon System (ATCRBS)? **B**
   A. An electronically steerable phased-array directional antenna.
   B. An L-band monopole blade-type omnidirectional antenna.
   C. A folded dipole reception antenna.
   D. An internally mounted, mechanically rotatable loop antenna.

Key Topic 73: Equipment Functions

1. Some aircraft and avionics equipment operates with a prime power line frequency of 400 Hz. What is the principle advantage of a higher line frequency? **C**
   A. 400 Hz power supplies draw less current than 60 Hz supplies allowing more current available for other systems on the aircraft.
   B. A 400 Hz power supply generates less heat and operates much more efficiently than a 60 Hz power supply.
   C. The magnetic devices in a 400 Hz power supply such as transformers, chokes and filters are smaller and lighter than those used in 60 Hz power supplies.
   D. 400 Hz power supplies are much less expensive to produce than power supplies with lower line frequencies.

2. Aviation services use predominantly ____ microphones. **A**
   A. Dynamic
   B. Carbon
   C. Condenser
   D. Piezoelectric crystal

3. Typical airborne HF transmitters usually provide a nominal RF power output to the antenna of ____ watts, compared with ____ watts RF output from a typical VHF transmitter. **D**
   A. 10, 50
   B. 50, 10
   C. 20, 100
   D. 100, 20

4. Before ground testing an aircraft RADAR, the operator should: **A**
   A. Ensure that the area in front of the antenna is clear of other maintenance personnel to avoid radiation hazards.
   B. Be sure the receiver has been properly shielded and grounded.
   C. First test the transmitter connected to a matched load.
   D. Measure power supply voltages to prevent circuit damage.

5. What type of antenna is used in an aircraft's Very High Frequency Omnidirectional Range (VOR) and Localizer (LOC) installations? **B**
   A. Vertically polarized antenna that radiates an omnidirectional antenna pattern.
   B. Horizontally polarized omnidirection reception antenna.
   C. Balanced loop transmission antenna.
   D. Folded dipole reception antenna.

6. What is the function of a commercial aircraft's SELCAL installation? SELCAL is a type of aircraft communications __________. **B**
   A. Device that allows an aircraft's receiver to be continuously calibrated for signal selectivity.
   B. System where a ground-based transmitter can call a selected aircraft or group of aircraft without the flight crew monitoring the ground-station frequency.
   C. Transmission that uses sequential logic algorithm encryption to prevent public "eavesdropping" of crucial aircraft flight data.
   D. System where an airborne transmitter can selectively calculate the line-of-sight distance to several ground-station receivers.

Subelement 3-L – Installation, Maintenance & Repair: 8 Key Topics - 8 Exam Questions

Key Topic 74: Indicating Meters

1. What is a 1/2 digit on a DMM? **B**

   A. Smaller physical readout on the left side of the display.
   B. Partial extended accuracy on lower part of the range.
   C. Smaller physical readout on the right side.
   D. Does not apply to DMMs.

2. A 50 microampere meter movement has an internal resistance of 2,000 ohms. What applied voltage is required to indicate half-scale deflection? **D**

   A. 0.01 volts
   B. 0.10 volts
   C. 0.005 volts.
   D. 0.05 volts.

3. What is the purpose of a series multiplier resistor used with a voltmeter? **A**

   A. It is used to increase the voltage-indicating range of the voltmeter.
   B. A multiplier resistor is not used with a voltmeter.
   C. It is used to decrease the voltage-indicating range of the voltmeter.
   D. It is used to increase the current-indicating range of an ammeter, not a voltmeter.

4. What is the purpose of a shunt resistor used with an ammeter? **C**

   A. A shunt resistor is not used with an ammeter.
   B, It is used to decrease the ampere indicating range of the ammeter.
   C. It is used to increase the ampere indicating range of the ammeter.
   D. It is used to increase the voltage indicating range of the voltmeter, not the ammeter.

5. What instrument is used to indicate high and low digital voltage states? **B**

   A. Ohmmeter.
   B. Logic probe.
   C. Megger.
   D. Signal strength meter.

6. What instrument may be used to verify proper radio antenna functioning? **C**

   A. Digital ohm meter.
   B. Hewlett-Packard frequency meter.
   C. An SWR meter.
   D. Different radio.

Key Topic 75: Test Equipment

1. How is a frequency counter used? **D**
   - A. To provide reference points on an analog receiver dial thereby aiding in the alignment of the receiver.
   - B. To heterodyne the frequency being measured with a known variable frequency oscillator until zero beat is achieved, thereby indicating the unknown frequency.
   - C. To measure the deviation in an FM transmitter in order to determine the percentage of modulation.
   - D. To measure the time between events, or the frequency, which is the reciprocal of the time.

2. What is a frequency standard? **B**
   - A. A well-known (standard) frequency used for transmitting certain messages.
   - B. A device used to produce a highly accurate reference frequency.
   - C. A device for accurately measuring frequency to within 1 Hz.
   - D. A device used to generate wide-band random frequencies.

3. What equipment may be useful to track down EMI aboard a ship or aircraft? **C**
   - A. Fluke multimeter.
   - B. An oscilloscope.
   - C. Portable AM receiver.
   - D. A logic probe.

4. On an analog wattmeter, what part of the scale is most accurate and how much does that accuracy extend to the rest of the reading scale? **A**
   - A. The accuracy is only at full scale, and that absolute number reading is carried through to the rest of the range. The upper 1/3 of the meter is the only truly calibrated part.
   - B. The accuracy is constant throughout the entire range of the meter.
   - C. The accuracy is only there at the upper 5% of the meter, and is not carried through at any other reading.
   - D. The accuracy cannot be determined at any reading.

5. Which of the following frequency standards is used as a time base standard by field technicians? **B**
   - A. Quartz Crystal.
   - B. Rubidium Standard.
   - C. Cesium Beam Standard.
   - D. LC Tank Oscillator.

6. Which of the following contains a multirange AF voltmeter calibrated in dB and a sharp, internal 1000 Hz bandstop filter, both used in conjunction with each other to perform quieting tests? **A**
   - A. SINAD meter.
   - B. Reflectometer.
   - C. Dip meter.
   - D. Vector-impedance meter.

Key Topic 76: Oscilloscopes

1. What is used to decrease circuit loading when using an oscilloscope? **B**
   - A. Dual input amplifiers.
   - B. 10:1 divider probe.
   - C. Inductive probe.
   - D. Resistive probe.

2. How does a spectrum analyzer differ from a conventional oscilloscope? **C**
   - A. The oscilloscope is used to display electrical signals while the spectrum analyzer is used to measure ionospheric reflection.
   - B. The oscilloscope is used to display electrical signals in the frequency domain while the spectrum analyzer is used to display electrical signals in the time domain.
   - C. The oscilloscope is used to display electrical signals in the time domain while the spectrum analyzer is used to display electrical signals in the frequency domain.
   - D. The oscilloscope is used for displaying audio frequencies and the spectrum analyzer is used for displaying radio frequencies.

3. What stage determines the maximum frequency response of an oscilloscope? **D**
   - A. Time base.
   - B. Horizontal sweep.
   - C. Power supply.
   - D. Vertical amplifier.

4. What factors limit the accuracy, frequency response, and stability of an oscilloscope? **A**
   - A. Sweep oscillator quality and deflection amplifier bandwidth.
   - B. Tube face voltage increments and deflection amplifier voltage.
   - C. Sweep oscillator quality and tube face voltage increments.
   - D. Deflection amplifier output impedance and tube face frequency increments.

5. An oscilloscope can be used to accomplish all of the following except: **C**
   - A. Measure electron flow with the aid of a resistor.
   - B. Measure phase difference between two signals.
   - C. Measure velocity of light with the aid of a light emitting diode.
   - D. Measure electrical voltage.

6. What instrument is used to check the signal quality of a single-sideband radio transmission? **D**
   - A. Field strength meter.
   - B. Signal level meter.
   - C. Sidetone monitor.
   - D. Oscilloscope.

Key Topic 77: Specialized Instruments

1. A(n) ____ and ____ can be combined to measure the characteristics of transmission lines. Such an arrangement is known as a time-domain reflectometer (TDR). **B**

   A. Frequency spectrum analyzer, RF generator.
   B. Oscilloscope, pulse generator.
   C. AC millivolt meter, AF generator.
   D. Frequency counter, linear detector.

2. What does the horizontal axis of a spectrum analyzer display? **D**

   A. Amplitude.
   B. Voltage.
   C. Resonance.
   D. Frequency.

3. What does the vertical axis of a spectrum analyzer display? **A**

   A. Amplitude.
   B. Duration.
   C. Frequency.
   D. Time.

4. What instrument is most accurate when checking antennas and transmission lines at the operating frequency of the antenna? **D**

   A. Time domain reflectometer.
   B. Wattmeter.
   C. DMM.
   D. Frequency domain reflectometer.

5. What test instrument can be used to display spurious signals in the output of a radio transmitter? **A**

   A. A spectrum analyzer.
   B. A wattmeter.
   C. A logic analyzer.
   D. A time domain reflectometer.

6. What instrument is commonly used by radio service technicians to monitor frequency, modulation, check receiver sensitivity, distortion, and to generate audio tones? **C**

   A. Oscilloscope.
   B. Spectrum analyzer.
   C. Service monitor.
   D. DMM.

Key Topic 78: Measurement Procedures

1. Can a P25 radio system be monitored with a scanner? **C**
   A. Yes , regardless if it has P25 decoding or not.
   B. No.
   C. Yes, if the scanner has P25 decoding.
   D. Yes, but it must also have P26 decoding.

2. Which of the following answers is true? **B**
   A. The RF Power reading on a CDMA (code division multiple access) radio will be very accurate on an analog power meter.
   B. The RF Power reading on a CDMA radio is not accurate on an analog power meter.
   C. Power cannot be measured using CDMA modulation.
   D. None of the above.

3. What is a common method used to program radios without using a "wired" connection? **C**
   A. Banding.
   B. Using the ultraviolet from a programmed radio to repeat the programming in another.
   C. Infra-red communication.
   D. Having the radio maker send down a programming signal via satellite.

4. What is the common method for determining the exact sensitivity specification of a receiver? **A**
   A. Measure the recovered audio for 12 dB of SINAD.
   B. Measure the recovered audio for 10 dB of quieting.
   C. Measure the recovered audio for 10 dB of SINAD.
   D. Measure the recovered audio for 25 dB of quieting.

5. A communications technician would perform a modulation-acceptance bandwidth test in order to: **D**
   A. Ascertain the audio frequency response of the receiver.
   B. Determine whether the CTCSS in the receiver is operating correctly.
   C. Verify the results from a 12 dB SINAD test.
   D. Determine the effective bandwidth of a communications receiver.

6. What is the maximum FM deviation for voice operation of a normal wideband channel on VHF and UHF? **B**
   A. 2.5 kHz
   B. 5.0 kHz
   C. 7.5 kHz
   D. 10 kHz

Key Topic 79: Repair Procedures

1. When soldering or working with CMOS electronics products or equipment, a wrist strap: **A**
   - A. Must have less than 100,000 ohms of resistance to prevent static electricity.
   - B. Cannot be used when repairing TTL devices.
   - C. Must be grounded to a water pipe.
   - D. Does not work well in conjunction with anti-static floor mats.

2. Which of the following is the preferred method of cleaning solder from plated-through circuit-board holes? **B**
   - A. Use a dental pick.
   - B. Use a vacuum device.
   - C. Use a soldering iron tip that has a temperature above 900 degrees F.
   - D. Use an air jet device.

3. What is the proper way to cut plastic wire ties? **D**
   - A. With scissors.
   - B. With a knife.
   - C. With semi-flush diagonal pliers.
   - D. With flush-cut diagonal pliers and cut flush.

4. The ideal method of removing insulation from wire is: **A**
   - A. The thermal stripper.
   - B. The pocket knife.
   - C. A mechanical wire stripper.
   - D. The scissor action stripping tool.

5. A "hot gas bonder" is used: **B**
   - A. To apply solder to the iron tip while it is heating the component.
   - B. For non-contact melting of solder.
   - C. To allow soldering both sides of the PC board simultaneously.
   - D. To cure LCA adhesives.

6. When repairing circuit board assemblies it is most important to: **C**
   - A. Use a dental pick to clear plated-through holes.
   - B. Bridge broken copper traces with solder.
   - C. Wear safety glasses.
   - D. Use a holding fixture.

Key Topic 80: Installation Codes & Procedures

1. What color is the binder for pairs 51-75 in a 100-pair cable? **D**

   A. Red
   B. Blue
   C. Black
   D. Green

2. What is most important when routing cables in a mobile unit? **C**

   A. That cables be cut to the exact length.
   B. Assuring accessibility of the radio for servicing from outside the vehicle.
   C. Assuring radio or electronics cables do not interfere with the normal operation of the vehicle.
   D. Assuring cables are concealed under floor mats or carpeting.

3. Why should you not use white or translucent plastic tie wraps on a radio tower? **B**

   A. White tie wraps are not FAA approved.
   B. UV radiation from the Sun deteriorates the plastic very quickly.
   C. The white color attracts wasps
   D. The black tie wraps may cause electrolysis.

4. What is the 6th pair color code in a 25 pair switchboard cable as is found in building telecommunications interconnections? **C**

   A. Blue/Green, Green/Blue.
   B. Red/Blue, White/Violet.
   C. Red/Blue, Blue/Red.
   D. White/Slate, Slate/White.

5. What tolerance off of plumb should a single base station radio rack be installed? **D**

   A. No tolerance allowed.
   B. Just outside the bubble on a level.
   C. All the way to one end.
   D. Just inside the bubble on a level.

6. What type of wire would connect an SSB automatic tuner to an insulated backstay? **A**

   A. GTO-15 high-voltage cable.
   B. RG8U.
   C. RG213.
   D. 16-gauge two-conductor.

Key Topic 81: Troubleshooting

1. On a 150 watt marine SSB HF transceiver, what would be indicated by a steady output of 75 watts when keying the transmitter on? **A**
   - A. There is probably a defect in the system causing the carrier to be transmitted.
   - B. One of the sidebands is missing.
   - C. Both sidebands are being transmitting.
   - D. The operation is normal.

2. The tachometer of a building's elevator circuit experiences interference caused by the radio system nearby. What is a common potential "fix" for the problem? **B**
   - A. Replace the tachometer of the elevator.
   - B. Add a .01 μF capacitor across the motor/tachometer leads.
   - C. Add a 200 μF capacity across the motor/tachometer leads.
   - D. Add an isolating resistor in series with the motor leads.

3. A common method of programming portable or mobile radios is to use a: **A**
   - A. A laptop computer.
   - B Dummy load.
   - C. A wattmeter.
   - D. A signal generator.

4. In a software-defined transceiver, what would be the best way for a technician to make a quick overall evaluation of the radio's operational condition? **C**
   - A. Set up a spectrum analyzer and service monitor and manually verify the manufacturer's specifications.
   - B. Use another radio on the same frequency to check the transmitter.
   - C. Use the built-in self-test feature.
   - D. Using on-board self-test routines are strictly prohibited by the FCC in commercial transmitters. Amateur Radio is the only service currently authorized to use them.

5. How might an installer verify correct GPS sentence to marine DSC VHF radio? **B**
   - A. Press and hold the red distress button.
   - B. Look for latitude and longitude on the display.
   - C. Look for GPS confirmation readout.
   - D. Ask for VHF radio check position report.

6. What steps must be taken to activate the DSC emergency signaling function on a marine VHF? **D**
   - A. Separate 12 volts to the switch.
   - B. Secondary DSC transmit antenna.
   - C. GPS position input.
   - D. Input of registered 9-digit MMSI.

Subelement 3-M – Communications Technology: 3 Key Topics - 3 Exam Questions

Key Topic 82: Types of Transmissions

1. What term describes a wide-bandwidth communications system in which the RF carrier frequency varies according to some predetermined sequence? **D**
   - A. Amplitude compandored single sideband.
   - B. SITOR.
   - C. Time-domain frequency modulation.
   - D. Spread spectrum communication.

2. Name two types of spread spectrum systems used in most RF communications applications? **C**
   - A. AM and FM.
   - B. QPSK or QAM.
   - C. Direct Sequence and Frequency Hopping.
   - D. Frequency Hopping and APSK.

3. What is the term used to describe a spread spectrum communications system where the center frequency of a conventional carrier is altered many times per second in accordance with a pseudo-random list of channels? **A**
   - A. Frequency hopping.
   - B. Direct sequence.
   - C. Time-domain frequency modulation.
   - D. Frequency compandored spread spectrum.

4. A TDMA radio uses what to carry the multiple conversations sequentially? **D**
   - A. Separate frequencies.
   - B. Separate pilot tones.
   - C. Separate power levels.
   - D. Separate time slots.

5. Which of the following statements about SSB voice transmissions is correct? **C**
   - A. They use A3E emission which are produced by modulating the final amplifier.
   - B. They use F3E emission which is produced by phase shifting the carrier.
   - C. They normally use J3E emissions, which consists of one sideband and a suppressed carrier.
   - D. They may use A1A emission to suppress the carrier.

6. What are the two most-used PCS (Personal Communications Systems) coding techniques used to separate different calls? **B**
   - A. QPSK and QAM.
   - B. CDMA and GSM.
   - C. ABCD and SYZ.
   - D. AM and Frequency Hopping.

Key Topic 83: Coding and Multiplexing

1. What is a CODEC? **C**

   A. A device to read Morse code.
   B. A computer operated digital encoding compandor.
   C. A coder/decoder IC or circuitry that converts a voice signal into a predetermined digital format for encrypted transmission.
   D. A voice amplitude compression chip.

2. The GSM (Global System for Mobile Communications) uses what type of CODEC for digital mobile radio system communications? **A**

   A. Regular-Pulse Excited (RPE).
   B. Code-Excited Linear Predictive (CLEP).
   C. Multi-Pulse Excited (MPE).
   D. Linear Excited Code (LEC).

3. Which of the following codes has gained the widest acceptance for exchange of data from one computer to another? **D**

   A. Gray.
   B. Baudot.
   C. Morse.
   D. ASCII.

4. The International Organization for Standardization has developed a seven-level reference model for a packet-radio communications structure. What level is responsible for the actual transmission of data and handshaking signals? **A**

   A. The physical layer.
   B. The transport layer.
   C. The communications layer.
   D. The synchronization layer.

5. What CODEC is used in Phase 2 P25 radios? **D**

   A. IWCE
   B. IMBC
   C. IMMM
   D. AMBE

6. The International Organization for Standardization has developed a seven-level reference model for a packet-radio communications structure. The _______ level arranges the bits frames and controls data flow. **B**

   A. Transport layer.
   B. Link layer.
   C. Communications layer.
   D. Synchronization layer.

Key Topic 84: Signal Processing, Software and Codes

1. What is a SDR? **B**
   A. Software Deviation Ratio.
   B. Software Defined Radio.
   C. SWR Meter.
   D. Static Dynamic Ram.

2. What does the DSP not do in a modern DSP radio? **D**
   A. Control frequency.
   B. Control modulation.
   C. Control detection.
   D. Control SWR.

3. Which statement best describes the code used for GMDSS-DSC transmissions? **A**
   A. A 10 bit error correcting code starting with bits of data followed by a 3 bit error correcting code.
   B. A 10 bit error correcting code starting with a 3 bit error correcting codefollowed by 7 bits of data.
   C. An 8 bit code with 7 bits of data followed by a single parity bit.
   D. A 7 bit code that is transmitted twice for error correction.

4. Which is the code used for SITOR-A and -B transmissions? **C**
   A. The 5 bit baudot telex code.
   B. Each character consists of 7 bits with 3 "zeros" and 4 "ones".
   C. Each character consists of 7 bits with 4 "zeros" and 3 "ones".
   D. Each character has 7 bits of data and 3 bits for error correction.

5. Which of the following statements is true? **B**
   A. The Signal Repetition character (1001100) is used as a control signal in SITOR-ARQ.
   B. The Idle Signal (a) (0000111) is used for FEC Phasing Signal 1.
   C. The Idle Signal (b) (0011001) is used for FEC Phasing Signal 2.
   D. The Control Signal 1 (0101100) is used to determine the time displacement in SITOR-B.

6. What principle allows multiple conversations to be able to share one radio channel on a GSM channel? **C**
   A. Frequency Division Multiplex.
   B. Double sideband.
   C. Time Division Multiplex.
   D. None of the above.

Subelement 3-N – Marine: 5 Key Topics - 5 Exam Questions

Key Topic 85: VHF

1. What is the channel spacing used for VHF marine radio? **D**
   - A. 10 kHz.
   - B. 12.5 kHz.
   - C. 20 kHz.
   - D. 25 kHz.

2. What VHF channel is assigned for distress and calling? **B**
   - A. 70
   - B. 16
   - C. 21A
   - D. 68

3. What VHF Channel is used for Digital Selective Calling and acknowledgement? **C**
   - A. 16
   - B. 21A
   - C. 70
   - D. 68

4. Maximum allowable frequency deviation for VHF marine radios is: **A**
   - A. +/- 5 kHz.
   - B. +/- 15 kHz.
   - C. +/- 2.5 kHz.
   - D. +/- 25 kHz.

5. What is the reason for the USA-INT control or function? **B**
   - A. It changes channels that are normally simplex channels into duplex channels.
   - B. It changes some channels that are normally duplex channels into simplex channels.
   - C. When the control is set to "INT" the range is increased.
   - D. None of the above.

6. How might an installer verify correct GPS sentence to marine DSC VHF radio? **A**
   - A. Look for latitude and longitude, plus speed, on VHF display.
   - B. Press and hold the red distress button.
   - C. Look for GPS confirmation readout.
   - D. Ask for VHF radio check position report.

Key Topic 86: MF-HF, SSB-SITOR

1. What is a common occurrence when voice-testing an SSB aboard a boat? **B**
   A. Ammeter fluctuates down with each spoken word.
   B. Voltage panel indicator lamps may glow with each syllable.
   C. Automatic tuner cycles on each syllable.
   D. Minimal voltage drop seen at power source.

2. What might contribute to apparent low voltage on marine SSB transmitting? **C**
   A. Blown red fuse.
   B. Too much grounding.
   C. Blown black negative fuse.
   D. Antenna mismatch.

3. What type of wire connects an SSB automatic tuner to an insulated backstay? **D**
   A. RG8U.
   B. RG213.
   C. 16-gauge two-conductor.
   D. GTO-15 high-voltage cable.

4. Which of the following statements concerning SITOR communications is true? **A**
   A. ARQ message transmissions are made in data groups consisting of three-character blocks.
   B. ARQ transmissions are acknowledged by the Information Receiving Station only at the end of the message.
   C. ARQ communications rely upon error correction by time diversity transmission and reception.
   D. Forward error correction is an interactive mode.

5. The sequence ARQ, FEC, SFEC best corresponds to which of the following sequences? **C**
   A. One-way communications to a single station, one-way communications to all stations, two-way communications.
   B. One-way communications to all stations, two-way communications, one-way communications to a single station.
   C. Two way communications, one-way communications to all stations, one-way communications to a single station.
   D. Two way communications, one-way communications to a single station, one-way communications to all stations.

6. Which of the following statements concerning SITOR communications is true? **D**
   A. Communication is established on the working channel and answerbacks are exchanged before FEC broadcasts can be received.
   B. In the ARQ mode each character is transmitted twice.
   C. Weather broadcasts cannot be made in FEC because sending each character twice would cause the broadcast to be prohibitively long.
   D. Two-way communication with the coast radio station using FEC is not necessaryto be able to receive the broadcasts.

Key Topic 87: Survival Craft Equipment: VHF, SARTs & EPIRBs

1. What causes the SART to begin a transmission? **B**
   - A. When activated manually, it begins radiating immediately.
   - B. After being activated the SART responds to RADAR interrogation.
   - C. It is either manually or water activated before radiating.
   - D. It begins radiating only when keyed by the operator.

2. How should the signal from a Search And Rescue RADAR Transponder appear on a RADAR display? **D**
   - A. A series of dashes.
   - B. A series of spirals all originating from the range and bearing of the SART.
   - C. A series of twenty dashes.
   - D. A series of 12 equally spaced dots.

3. In which frequency band does a search and rescue transponder operate? **A**
   - A. 9 GHz
   - B. 3 GHz
   - C. S-band
   - D. 406 MHz

4. Which piece of required GMDSS equipment is the primary source of transmitting locating signals? **D**
   - A. Radio Direction Finder (RDF).
   - B. A SART transmitting on 406 MHz.
   - C. Survival Craft Transceiver.
   - D. An EPIRB transmitting on 406 MHz.

5. Which of the following statements concerning satellite EPIRBs is true? **A**
   - A. Once activated, these EPIRBs transmit a signal for use in identifying the vessel and for determining the position of the beacon.
   - B. The coded signal identifies the nature of the distress situation.
   - C. The coded signal only identifies the vessel's name and port of registry.
   - D. If the GMDSS Radio Operator does not program the EPIRB, it will transmit default information such as the follow-on communications frequency and mode.

6. What statement is true regarding 406 MHz EPIRB transmissions? **C**
   - A. Allows immediate voice communications with the RCC.
   - B. Coding permits the SAR authorities to know if manually or automatically activated.
   - C. Transmits a unique hexadecimal identification number.
   - D. Radio Operator programs an I.D. into the SART immediately prior to activation.

Key Topic 88: FAX, NAVTEX

1. What is facsimile? **C**

   A. The transmission of still pictures by slow-scan television.
   B. The transmission of characters by radioteletype that form a picture when printed.
   C. The transmission of printed pictures for permanent display on paper.
   D. The transmission of video by television.

2. What is the standard scan rate for high-frequency 3 MHz - 23 MHz weather facsimile reception from shore stations? **B**

   A. 240 lines per minute.
   B. 120 lines per minute.
   C. 150 lines per second.
   D. 60 lines per second.

3. What would be the bandwidth of a good crystal lattice band-pass filter for weather facsimile HF (high frequency) reception? **C**

   A. 500 Hz at -6 dB.
   B. 6 kHz at -6 dB.
   C. 1 kHz at -6 dB.
   D. 15 kHz at -6 dB.

4. Which of the following statements about NAVTEX is true? **A**

   A. Receives MSI broadcasts using SITOR-B or FEC mode.
   B. The ship station transmits on 518 kHz.
   C. The ship receives MSI broadcasts using SITOR-A or ARQ mode.
   D. NAVTEX is received on 2182 kHz using SSB.

5. Which of the following is the primary frequency that is used exclusively for NAVTEX broadcasts internationally? **D**

   A. 2187.5 kHz.
   B. 4209.5 kHz.
   C. VHF channel 16.
   D. 518 kHz.

6. What determines whether a NAVTEX receiver does not print a particular type of message content? **B**

   A. The message does not concern your vessel.
   B. The subject indicator matches that programmed for rejection by the operator.
   C. The transmitting station ID covering your area has not been programmed for rejection by the operator.
   D. All messages sent during each broadcast are printed.

Key Topic 89: NMEA Data

1. What data language is bi-directional, multi-transmitter, multi-receiver network? **A**
   - A. NMEA 2000.
   - B. NMEA 0181.
   - C. NMEA 0182.
   - D. NMEA 0183.

2. How should shielding be grounded on an NMEA 0183 data line? **B**
   - A. Unterminated at both ends.
   - B. Terminated to ground at the talker and unterminated at the listener.
   - C. Unterminated at the talker and terminated at the listener.
   - D. Terminated at both the talker and listener.

3. What might occur in NMEA 2000 network topology if one device in line should fail? **D**
   - A. The system shuts down until the device is removed.
   - B. Other electronics after the failed device will be inoperable.
   - C. The main fuse on the backbone may open.
   - D. There will be no interruption to all other devices.

4. In an NMEA 2000 device, a load equivalence number (LEN) of 1 is equivalent to how much current consumption? **A**
   - A. 50 mA
   - B. 10 mA
   - C. 25 mA
   - D. 5 mA

5. An NMEA 2000 system with devices in a single location may be powered using this method: **B**
   - A. Dual mid-powered network.
   - B. End-powered network.
   - C. Individual devices individually powered.
   - D. No 12 volts needed for NMEA 2000 devices.

6. What voltage drop at the end of the last segment will satisfy NMEA 2000 network cabling plans? **C**
   - A. 0.5 volts
   - B. 2.0 volts
   - C. 1.5 volts
   - D. 3.0 volts

Subelement 3-O – RADAR: 5 Key Topics - 5 Exam Questions

Key Topic 90: RADAR Theory

1 What is the normal range of pulse repetition rates? **D**

A. 2,000 to 4,000 pps.
B. 1,000 to 3,000 pps.
C. 500 to 1,000 pps.
D. 500 to 2,000 pps.

2. The RADAR range in nautical miles to an object can be found by measuring the elapsed time during a RADAR pulse and dividing this quantity by: **C**

A. 0.87 seconds.
B. 1.15 µs.
C. 12.346 µs.
D. 1.073 µs.

3. What is the normal range of pulse widths? **B**

A. .05 µs to 0.1 µs.
B. .05 µs to 1.0 µs.
C. 1.0 µs to 3.5 µs.
D. 2.5 µs to 5.0 µs.

4. Shipboard RADAR is most commonly operated in what band? **C**

A. VHF.
B. UHF.
C. SHF.
D. EHF.

5. The pulse repetition rate (prr) of a RADAR refers to the: **D**

A. Reciprocal of the duty cycle.
B. Pulse rate of the local oscillator.
C. Pulse rate of the klystron.
D. Pulse rate of the magnetron.

6. If the elapsed time for a RADAR echo is 62 microseconds, what is the distance in nautical miles to the object? **A**

A. 5 nautical miles.
B. 87 nautical miles.
C. 37 nautical miles.
D. 11.5 nautical miles.

Key Topic 91: Components

1. The ATR box: **A**

   A. Prevents the received signal from entering the transmitter.
   B. Protects the receiver from strong RADAR signals.
   C. Turns off the receiver when the transmitter is on.
   D. All of the above.

2. What is the purpose or function of the RADAR duplexer/circulator? It is a/an: **B**

   A. Coupling device that is used in the transition from a rectangular waveguide to a circular waveguide.
   B. Electronic switch that allows the use of one antenna for both transmission and reception.
   C. Modified length of waveguide that is used to sample a portion of the transmitted energy for testing purposes.
   D. Dual section coupling device that allows the use of a magnetron as a transmitter.

3. What device can be used to determine the performance of a RADAR system at sea? **A**

   A. Echo box.
   B. Klystron.
   C. Circulator.
   D. Digital signal processor.

4. What is the purpose of a synchro transmitter and receiver? **C**

   A. Synchronizes the transmitted and received pulse trains.
   B. Prevents the receiver from operating during the period of the transmitted pulse.
   C. Transmits the angular position of the antenna to the indicator unit.
   D. Keeps the speed of the motor generator constant.

5. Digital signal processing (DSP) of RADAR signals (compared with analog) causes: **B**

   A. Improved display graphics.
   B. Improved weak signal or target enhancement.
   C. Less interference with SONAR systems.
   D. Less interference with other radio communications equipment.

6. The component or circuit providing the transmitter output power for a RADAR system is the: **D**

   A. Thyratron.
   B. SCR.
   C. Klystron.
   D. Magnetron.

Key Topic 92: Range, Pulse Width & Repetition Rate

1. When a RADAR is being operated on the 48 mile range setting, what is the most appropriate pulse width (PW) and pulse repetition rate (pps)? **D**
   A. 1.0 μs PW and 2,000 pps.
   B. 0.05 μs PW and 2,000 pps.
   C. 2.5 μs PW and 2,500 pps.
   D. 1.0 μs PW and 500 pps.

2. When a RADAR is being operated on the 6 mile range setting what is the most appropriate pulse width and pulse repetition rate? **C**
   A. 1.0 μs PW and 500 pps.
   B. 2.0 μs PW and 3,000 pps.
   C. 0.25 μs PW and 1,000 pps.
   D. 0.01 μs PW and 500 pps.

3. We are looking at a target 25 miles away. When a RADAR is being operated on the 25 mile range setting what is the most appropriate pulse width and pulse repetition rate? **A**
   A. 1.0 μs PW and 500 pps.
   B. 0.25 μs PW and 1,000 pps.
   C. 0.01 μs PW and 500 pps.
   D. 0.05 μs PW and 2,000 pps.

4. What pulse width and repetition rate should you use at long ranges? **D**
   A. Narrow pulse width and slow repetition rate.
   B. Narrow pulse width and fast repetition rate.
   C. Wide pulse width and fast repetition rate.
   D. Wide pulse width and slow repetition rate.

5. What pulse width and repetition rate should you use at short ranges? **C**
   A. Wide pulse width and fast repetition rate.
   B. Narrow pulse width and slow repetition rate.
   C. Narrow pulse width and fast repetition rates.
   D. Wide pulse width and slow repetition rates.

6. When a RADAR is being operated on the 1.5 mile range setting, what is the most appropriate pulse width and pulse repetition rate? **B**
   A. 0.25 μs PW and 1,000 pps.
   B. 0.05 μs PW and 2,000 pps.
   C. 1.0 μs PW and 500 pps.
   D. 2.5 μs PW and 2,500 pps.

Key Topic 93: Antennas & Waveguides

1. How does the gain of a parabolic dish antenna change when the operating frequency is doubled? **C**

   A. Gain does not change.
   B. Gain is multiplied by 0.707.
   C. Gain increases 6 dB.
   D. Gain increases 3 dB.

2. What type of antenna or pickup device is used to extract the RADAR signal from the wave guide? **A**

   A. J-hook.
   B. K-hook.
   C. Folded dipole.
   D. Circulator.

3. What happens to the beamwidth of an antenna as the gain is increased? The beamwidth: **D**

   A. Increases geometrically as the gain is increased.
   B. Increases arithmetically as the gain is increased.
   C. Is essentially unaffected by the gain of the antenna.
   D. Decreases as the gain is increased.

4. A common shipboard RADAR antenna is the: **A**

   A. Slotted array.
   B. Dipole.
   C. Stacked Yagi.
   D. Vertical Marconi.

5. Conductance takes place in a waveguide: **D**

   A. By interelectron delay.
   B. Through electrostatic field reluctance.
   C. In the same manner as a transmission line.
   D. Through electromagnetic and electrostatic fields in the walls of the waveguide.

6. To couple energy into and out of a waveguide use: **B**

   A. Wide copper sheeting.
   B. A thin piece of wire as an antenna.
   C. An LC circuit.
   D. Capacitive coupling.

Key Topic 94: RADAR Equipment

1. The permanent magnetic field that surrounds a traveling-wave tube (TWT) is intended to: **B**

   A. Provide a means of coupling.
   B. Prevent the electron beam from spreading.
   C. Prevent oscillations.
   D. Prevent spurious oscillations.

2. Prior to testing any RADAR system, the operator should first: **D**

   A. Check the system grounds.
   B. Assure the display unit is operating normally.
   C. Inform the airport control tower or ship's master.
   D. Assure no personnel are in front of the antenna.

3. In the term "ARPA RADAR," ARPA is the acronym for which of the following? **A**

   A. Automatic RADAR Plotting Aid.
   B. Automatic RADAR Positioning Angle.
   C. American RADAR Programmers Association.
   D. Authorized RADAR Programmer and Administrator.

4. Which of the following is NOT a precaution that should be taken to ensure the magnetron is not weakened: **C**

   A. Keep metal tools away from the magnet.
   B. Do not subject it to excessive heat.
   C. Keep the TR properly tuned.
   D. Do not subject it to shocks and blows.

5. Exposure to microwave energy from RADAR or other electronics devices is limited by U.S Health Department regulations to _______ mW/centimeter. **B**

   A. 0.005
   B. 5.0
   C. 0.05
   D. 0.5

6. RADAR collision avoidance systems utilize inputs from each of the following except your ship's: **C**

   A. Gyrocompass.
   B. Navigation position receiver.
   C. Anemometer.
   D. Speed indicator.

Subelement 3-P – Satellite: 4 Key Topics - 4 Exam Questions

Key Topic 95: Low Earth Orbit Systems

1. What is the orbiting altitude of the Iridium satellite communications system? **D**
   - A. 22,184 miles.
   - B. 11,492 miles.
   - C. 4,686 miles.
   - D. 485 miles.

2. What frequency band is used by the Iridium system for telephone and messaging? **B**
   - A. 965 - 985 MHz.
   - B. 1616 -1626 MHz.
   - C. 1855 -1895 MHz.
   - D. 2415 - 2435 MHz.

3. What services are provided by the Iridium system? **C**
   - A. Analog voice and Data at 4.8 kbps.
   - B. Digital voice and Data at 9.6 kbps.
   - C. Digital voice and Data at 2.4 kbps.
   - D. Analog voice and Data at 9.6 kbps.

4. Which of the following statements about the Iridium system is true? **A**
   - A. There are 48 spot beams per satellite with a footprint of 30 miles in diameter.
   - B. There are 48 satellites in orbit in 4 orbital planes.
   - C. The inclination of the orbital planes is 55 degrees.
   - D. The orbital period is approximately 85 minutes.

5. What is the main function of the COSPAS-SARSAT satellite system? **B**
   - A. Monitor 121.5 MHz for voice distress calls.
   - B. Monitor 406 MHz for distress calls from EPIRBs.
   - C. Monitor 1635 MHz for coded distress calls.
   - D. Monitor 2197.5 kHz for hexadecimal coded DSC distress messages.

6. How does the COSPAS-SARSAT satellite system determine the position of a ship in distress? **A**
   - A. By measuring the Doppler shift of the 406 MHz signal taken at several differentpoints in its orbit.
   - B. The EPIRB always transmits its position which is relayed by the satellite to the Local User Terminal.
   - C. It takes two different satellites to establish an accurate position.
   - D. None of the above.

Key Topic 96: INMARSAT Communications Systems-1

1. What is the orbital altitude of INMARSAT Satellites? **B**
   - A. 16, 436 miles.
   - B. 22,177 miles.
   - C. 10, 450 miles.
   - D. 26,435 miles.

2. Which of the following describes the INMARSAT Satellite system? **C**
   - A. AOR at 35° W, POR-E at 165° W, POR-W at 155° E and IOR at 56.5° E.
   - B. AOR-E at 25° W, AOR-W at 85° W, POR at 175° W and IOR at 56.5° E.
   - C. AOR-E at 15.5° W, AOR-W at 54° W, POR at 178° E and IOR at 64.5° E.
   - D. AOR at 40° W, POR at 178° W, IOR-E at 109° E and IOR-W at 46° E.

3. What are the directional characteristics of the INMARSAT-C SES antenna? **D**
   - A. Highly directional parabolic antenna requiring stabilization.
   - B. Wide beam width in a cardioid pattern off the front of the antenna.
   - C. Very narrow beam width straight-up from the top of the antenna.
   - D. Omnidirectional.

4. When engaging in voice communications via an INMARSAT-B terminal, what techniques are used? **A**
   - A. CODECs are used to digitize the voice signal.
   - B. Noise-blanking must be selected by the operator.
   - C. The voice signal must be compressed to fit into the allowed bandwidth.
   - D. The voice signal will be expanded at the receiving terminal.

5. Which of the following statements concerning INMARSAT geostationary satellites is true? **C**
   - A. They are in a polar orbit, in order to provide true global coverage.
   - B. They are in an equatorial orbit, in order to provide true global coverage.
   - C. They provide coverage to vessels in nearly all of the world's navigable waters.
   - D. Vessels sailing in equatorial waters are able to use only one satellite, whereas other vessels are able to choose between at least two satellites.

6. Which of the following conditions can render INMARSAT -B communications impossible? **D**
   - A. An obstruction, such as a mast, causing disruption of the signal between the satellite and the SES antenna when the vessel is steering a certain course.
   - B. A satellite whose signal is on a low elevation, below the horizon.
   - C. Travel beyond the effective radius of the satellite.
   - D. All of these.

Key Topic 97: INMARSAT Communications Systems-2

1. What is the best description for the INMARSAT-C system? **B**

   A. It provides slow speed telex and voice service.
   B. It is a store-and-forward system that provides routine and distress communications.
   C. It is a real-time telex system.
   D. It provides world-wide coverage.

2. The INMARSAT mini-M system is a: **D**

   A. Marine SONAR system.
   B. Marine global satellite system.
   C. Marine depth finder.
   D. Satellite system utilizing spot beams to provide for small craft communications.

3. What statement best describes the INMARSAT-B services? **A**

   A. Voice at 16 kbps, Fax at 14.4 kbps and high-speed Data at 64/54.
   B. Store and forward high speed data at 36/48 kbps.
   C. Voice at 3 kHz, Fax at 9.6 kbps and Data at 4.8 kbps.
   D. Service is available only in areas served by highly directional spot beam antennas.

4. Which INMARSAT systems offer High Speed Data at 64/54 kbps? **D**

   A. C.
   B. B and C.
   C. Mini-M.
   D. B, M4 and Fleet.

5. When INMARSAT-B and INMARSAT-C terminals are compared: **A**

   A. INMARSAT-C antennas are small and omni-directional, while INMARSAT-B antennasare larger and directional.
   B. INMARSAT-B antennas are bulkier but omni-directional, while INMARSAT-C antennasare smaller and parabolic, for aiming at the satellite.
   C. INMARSAT-B antennas are parabolic and smaller for higher gain, while INMARSAT-C antennas are larger but omni-directional.
   D. INMARSAT-C antennas are smaller but omni-directional, while I INMARSAT-B antennas are parabolic for lower gain.

6. What services are provided by the INMARSAT-M service? **C**

   A. Data and Fax at 4.8 kbps plus e-mail.
   B. Voice at 3 kHz, Fax at 9.6 kbps and Data at 4.8 kbps.
   C. Voice at 6.2 kbps, Data at 2.4 kbps, Fax at 2.4 kbps and e-mail.
   D. Data at 4.8 kbps and Fax at 9.6 kbps plus e-mail.

Key Topic 98: GPS

1. Global Positioning Service (GPS) satellite orbiting altitude is: **C**

   A. 4,686 miles.
   B. 24,184 miles.
   C. 12,554 miles.
   D. 247 miles.

2. The GPS transmitted frequencies are: **B**

   A. 1626.5 MHz and 1644.5 MHz.
   B. 1227.6 MHz and 1575.4 MHz.
   C. 2245.4 and 2635.4 MHz.
   D. 946.2 MHz and 1226.6 MHz.

3. How many GPS satellites are normally in operation? **C**

   A. 8
   B. 18
   C. 24
   D. 36

4. What best describes the GPS Satellites orbits? **A**

   A. They are in six orbital planes equally spaced and inclined about 55 degrees to the equator.
   B. They are in four orbital planes spaced 90 degrees in a polar orbit.
   C. They are in a geosynchronous orbit equally spaced around the equator.
   D. They are in eight orbital planes at an altitude of approximately 1,000 miles.

5. How many satellites must be received to provide complete position and time? **D**

   A. 1
   B. 2
   C. 3
   D. 4

6. What is DGPS? **B**

   A. Digital Ground Position System.
   B. A system to provide additional correction factors to improve position accuracy.
   C. Correction signals transmitted by satellite.
   D. A system for providing altitude corrections for aircraft.

Subelement 3-Q – SAFETY: 2 Key Topics - 2 Exam Questions

Key Topic 99: Radiation Exposure

1. Compliance with MPE, or Maximum Permissible Exposure to RF levels (as defined in FCC Part 1, OET Bulletin 65) for "controlled" environments, are averaged over _______ minutes, while "uncontrolled" RF environments are averaged over ______ minutes. **A**
   - A. 6, 30.
   - B. 30, 6.
   - C. 1, 15.
   - D. 15, 1.

2. Sites having multiple transmitting antennas must include antennas with more than _______% of the maximum permissible power density exposure limit when evaluating RF site exposure. **B**
   - A. Any
   - B. 5
   - C. 1
   - D. 12.5

3. RF exposure from portable radio transceivers may be harmful to the eyes because: **D**
   - A. Magnetic fields blur vision.
   - B. RF heating polarizes the eye lens.
   - C. The magnetic field may attract metal particles to the eye.
   - D. RF heating may cause cataracts.

4. At what aggregate power level is an MPE (Maximum Permissible Exposure) study required? **A**
   - A. 1000 Watts ERP.
   - B. 500 Watts ERP.
   - C. 100 Watts ERP.
   - D. Not required.

5. Why must you never look directly into a fiber optic cable? **B**
   - A. High power light waves can burn the skin surrounding the eye.
   - B. An active fiber signal may burn the retina and infra-red light cannot be seen.
   - C. The end is easy to break.
   - D. The signal is red and you can see it.

6. If the MPE (Maximum Permissible Exposure) power is present, how often must the personnel accessing the affected area be trained and certified? **C**
   - A. Weekly.
   - B. Monthly.
   - C. Yearly.
   - D. Not at all.

Key Topic 100: Safety Steps

1. What device can protect a transmitting station from a direct lightning hit? **D**

   A. Lightning protector.
   B. Grounded cabinet.
   C. Short lead in.
   D. There is no device to protect a station from a direct hit from lightning.

2. What is the purpose of not putting sharp corners on the ground leads within a building? **C**

   A. No reason.
   B. It is easier to install.
   C. Lightning will jump off of the ground lead because it is not able to make sharp bends.
   D. Ground leads should always be made to look good in an installation, including the use of sharp bends in the corners.

3. Should you use a power drill without eye protection? **B**

   A. Yes.
   B. No.
   C. It's okay as long as you keep your face away from the drill bit.
   D. Only in an extreme emergency.

4. What class of fire is one that is caused by an electrical short circuit and what is the preferred substance used to extinguish that type of fire? **C**

   A. FE28.
   B. FE29.
   C. FE30.
   D. FE31.

5. Do shorted-stub lightning protectors work at all frequencies? **D**

   A. Yes.
   B. No, the short also kills the radio signals.
   C. No, the short enhances the radio signal at the tuned band.
   D. No, only at the tuned frequency band.

6. What is a GFI electrical socket used for? **A**

   A. To prevent electrical shock by sensing ground path current and shutting the circuit down.
   B. As a gold plated socket.
   C. To prevent children from sticking objects in the socket.
   D. To increase the current capacity of the socket.

RADAR Endorsement
Element 8 Questions

The Element 8 Radar Endorsement is not a separate FCC License. You must have passed the GROP Element 3 exam successfully prior to obtaining the Element 8 Special Endorsement.

This endorsement is added to a GROP and authorizes the license holder to legally adjust, repair, install, service and maintain aviation and ship RADAR navigation equipment.

The 50 questions asked on your exam will be taken for an FCC question pool of 300 questions covering six sub-element question groups. Within these groups are a grand total of 50 "Key Topics".

Your Element 8 exam will consist of one randomly selected multiple-choice question from each of the 50 "Key Topics" for a complete examination. The minimum passing score is 38 of the multiple-choice questions answered correctly. Almost every page has a complete Key Topic of six questions and you will see one of these on your exam.

Because of the highly technical natural of the Element 8 exam, it is not recommended to study for this endorsement test at the same time you are learning material in the other elements. Of course, if you later plan to attempt taking this endorsement exam, it may be a good idea to sit for this test and just guess the answers to size up the Element 8 exam experience prior to studying for it in the future.

Contents of the Element 8 Examination:

Subelement A - RADAR Principles

Marine RADAR Systems - Key Topic 1
Distance and Time - Key Topic 2
Frequency and Wavelength - Key Topic 3
Power, Pulse Width, PRR - Key Topic 4
Range, Pulse Width, PRF - Key Topic 5
Pulse Width and Pulse Repetition Rates - Key Topic 6
Components-1 - Key Topic 7
Components-2 - Key Topic 8
Circuits-1 - Key Topic 9
Circuits-2 - Key Topic 10

Subelement B - Transmitting Systems

Transmitting Systems - Key Topic 11
Magnetrons - Key Topic 12
Modulation - Key Topic 13

Pulse Forming Networks Modulation - Key Topic 14
TR, ATR, Circulators and Directional Couplers-1 - Key Topic 15
TR, ATR, Circulators and Directional Couplers-2 - Key Topic 16
Timer- Trigger- Synchronizer Circuits - Key Topic 17
Power Supplies - Key Topic 18

Subelement C - Receiving Systems

Receiving Systems - Key Topic 19
Mixers - Key Topic 20
Local Oscillators - Key Topic 21
Amplifiers - Key Topic 22
Detectors and Video Amplifiers - Key Topic 23
Automatic Frequency Control - Key Topic 24
Sea Clutter - Key Topic 25
Power Supplies - Key Topic 26
Interference Issues - Key Topic 27
Miscellaneous - Key Topic 28

Subelement D - Display and Control Systems

Displays - Key Topic 29
Video Amplifiers and Sweep Circuits - Key Topic 30
Timing Circuits - Key Topic 31
Fixed Range Markers - Key Topic 32
Variable Range Markers - Key Topic 33
EBL, Azimuth and True Bearing - Key Topic 34
Memory Systems - Key Topic 35
ARPA, CAS - Key Topic 36
Display Systems Power Supplies - Key Topic 37
Miscellaneous - Key Topic 38

Subelement E - Antenna Systems

Antenna Systems - Key Topic 39
Transmission Lines - Key Topic 40
Antenna to display Interface - Key Topic 41
Wave Guides-1 - Key Topic 42
Wave Guides-2 - Key Topic 43

Subelement F - Installation, Maintenance and Repair

Equipment Faults-1 - Key Topic 44
Equipment Faults-2 - Key Topic 45
Equipment Faults-3 - Key Topic 46
Equipment Faults-4 - Key Topic 47
Maintenance - Key Topic 48
Installation - Key Topic 49
Safety - Key Topic 50

FCC Commercial Element 8

Subelement A - RADAR Principles: 10 Key Topics - 10 Exam Questions

Key Topic 1 - Marine RADAR Systems

1. Choose the most correct statement containing the parameters which control the size of the target echo. **A**
   - A. Transmitted power, antenna effective area, transmit and receive losses, RADAR cross section of the target, range to target.
   - B. Height of antenna, power radiated, size of target, receiver gain, pulse width.
   - C. Power radiated, antenna gain, size of target, shape of target, pulse width, receiver gain.
   - D. Magnetron gain, antenna gain, size of target, range to target, wave-guide loss.

2. Which of the following has NO effect on the maximum range capability? **B**
   - A. Carrier frequency.
   - B. Recovery time.
   - C. Pulse repetition frequency.
   - D. Receiver sensitivity.

3. What type of transmitter power is measured over a period of time? **A**
   - A. Average.
   - B. Peak.
   - C. Reciprocal.
   - D. Return.

4. What RADAR component controls timing throughout the system? **C**
   - A. Power supply.
   - B. Indicator.
   - C. Synchronizer.
   - D. Receiver.

5. Which of the following components allows the use of a single antenna for both transmitting and receiving? **B**
   - A. Mixer.
   - B. Duplexer.
   - C. Synchronizer.
   - D. Modulator.

6. The sweep frequency of a RADAR indicator is determined by what parameter? **D**
   - A. Carrier frequency.
   - B. Pulse width.
   - C. Duty cycle.
   - D. Pulse repetition frequency.

Key Topic 2 - Distance and Time

1. A radio wave will travel a distance of three nautical miles in: **D**

   A. 6.17 microseconds.
   B. 37.0 microseconds.
   C. 22.76 microseconds.
   D. 18.51 microseconds.

2.One RADAR mile is how many microseconds? **C**

   A. 6.2
   B. 528.0
   C. 12.34
   D. 0.186

3. RADAR range is measured by the constant: **A**

   A. 150 meters per microsecond.
   B. 150 yards per microsecond.
   C. 300 yards per microsecond.
   D. 18.6 miles per microsecond.

4. If a target is 5 miles away, how long does it take for the RADAR echo to be received back at the antenna? **D**

   A. 51.4 microseconds.
   B. 123 microseconds.
   C. 30.75 microseconds.
   D. 61.7 microseconds.

5. How long would it take for a RADAR pulse to travel to a target 10 nautical miles away and return to the RADAR receiver? **C**

   A. 12.34 microseconds.
   B. 1.234 microseconds.
   C. 123.4 microseconds.
   D. 10 microseconds.

6. What is the distance in nautical miles to a target if it takes 308.5 microseconds for the RADAR pulse to travel from the RADAR antenna to the target and back. **B**

   A. 12.5 nautical miles.
   B. 25 nautical miles.
   C. 50 nautical miles.
   D. 2.5 nautical miles.

Key Topic 3 - Frequency and Wavelength

1. Frequencies generally used for marine RADAR are in the ___ part of the radio spectrum. **C**

   A. UHF
   B. EHF
   C. SHF
   D. VHF

2. Practical RADAR operation requires the use of microwave frequencies so that: **A**

   A. Stronger target echoes will be produced.
   B. Ground clutter interference will be minimized.
   C. Interference to other communication systems will be eliminated.
   D. Non-directional antennas can be used for both transmitting and receiving.

3. An S-band RADAR operates in which frequency band? **D**

   A. 1 - 2 GHz.
   B. 4 - 8 GHz.
   C. 8 - 12 GHz.
   D. 2 - 4 GHz.

4. A RADAR operating at a frequency of 3 GHz has a wavelength of approximately: **B**

   A. 1 centimeter.
   B. 10 centimeters.
   C. 3 centimeters.
   D. 30 centimeters.

5. The major advantage of an S-band RADAR over an X-band RADAR is: **A**

   A. It is less affected by weather conditions.
   B. It has greater bearing resolution.
   C. It is mechanically less complex.
   D. It has greater power output.

6. An X band RADAR operates in which frequency band? **D**

   A. 1 - 2 GHz.
   B. 2 - 4 GHz.
   C. 4 - 8 GHz.
   D. 8 - 12 GHz.

Key Topic 4 - Power, Pulse Width, PRR

1. A pulse RADAR has a pulse repetition frequency (PRF) of 400 Hz, a pulse width of 1 microsecond, and a peak power of 100 kilowatts. The average power of the RADAR transmitter is: **B**

   A. 25 watts.
   B. 40 watts.
   C. 250 watts.
   D. 400 watts.

2. A shipboard RADAR transmitter has a pulse repetition frequency (PRF) of 1,000 Hz, a pulse width of 0.5 microseconds, peak power of 150 KW, and a minimum range of 75 meters. Its duty cycle is: **D**

   A. 0.5
   B. 0.05
   C. 0.005
   D. 0.0005

3. A pulse RADAR transmits a 0.5 microsecond RF pulse with a peak power of 100 kilowatts every 1600 microseconds. This RADAR has: **A**

   A. An average power of 31.25 watts.
   B. A PRF of 3,200.
   C. A maximum range of 480 kilometers.
   D. A duty cycle of 3.125 percent.

4. If a RADAR transmitter has a pulse repetition frequency (PRF) of 900 Hz, a pulse width of 0.5 microseconds and a peak power of 15 kilowatts, what is its average power output? **C**

   A. 15 kilowatts.
   B. 13.5 watts.
   C. 6.75 watts.
   D. 166.67 watts.

5. What is the average power if the RADAR set has a PRF of 1000 Hz, a pulse width of 1 microsecond, and a peak power rating of 100 kilowatts? **B**

   A. 10 watts.
   B. 100 watts.
   C. 1,000 watts.
   D. None of these.

6. A search RADAR has a pulse width of 1.0 microsecond, a pulse repetition frequency (PRF) of 900 Hz, and an average power of 18 watts. The unit's peak power is: **C**

   A. 200 kilowatts.
   B. 180 kilowatts.
   C. 20 kilowatts.
   D. 2 kilowatts.

Key Topic 5 - Range, Pulse Width, PRF

1. For a range of 5 nautical miles, the RADAR pulse repetition frequency should be: **D**

   A. 16.2 Hz or more.
   B. 16.2 MHz or less.
   C. 1.62 kHz or more.
   D. 16.2 kHz or less.

2. For a range of 100 nautical miles, the RADAR pulse repetition frequency should be: **B**

   A. 8.1 kHz or less.
   B. 810 Hz or less.
   C. 8.1 kHz or more.
   D. 81 kHz or more.

3. The minimum range of a RADAR is determined by: **C**

   A. The frequency of the RADAR transmitter.
   B. The pulse repetition rate.
   C. The transmitted pulse width.
   D. The pulse repetition frequency.

4. Short range RADARs would most likely transmit: **A**

   A. Narrow pulses at a fast rate.
   B. Narrow pulses at a slow rate.
   C. Wide pulses at a fast rate.
   D. Wide pulses at a slow rate.

5. For a range of 30 nautical miles, the RADAR pulse repetition frequency should be: **B**

   A. 0.27 kHz or less.
   B. 2.7 kHz or less.
   C. 27 kHz or more.
   D. 2.7 Hz or more.

6. For a range of 10 nautical miles, the RADAR pulse repetition frequency (PRF) should be **A**

   A. Approximately 8.1 kHz or less.
   B. 900 Hz.
   C. 18.1 kHz or more.
   D. 120.3 microseconds.

Key Topic 6: Pulse Width - Pulse Repetition Rates

1. If the PRF is 2500 Hz, what is the PRI? **B**

   A. 40 microseconds.
   B. 400 microseconds.
   C. 250 microseconds.
   D. 800 microseconds.

2. If the pulse repetition frequency (PRF) is 2000 Hz, what is the pulse repetition interval (PRI)? **C**

   A. 0.05 seconds.
   B. 0.005 seconds.
   C. 0.0005 seconds.
   D. 0.00005 seconds.

3. The pulse repetition rate (PRR) refers to: **D**

   A. The reciprocal of the duty cycle.
   B. The pulse rate of the local oscillator tube.
   C. The pulse rate of the klystron.
   D. The pulse rate of the magnetron.

4. If the RADAR unit has a pulse repetition frequency (PRF) of 2000 Hz and a pulse width of 0.05 microseconds, what is the duty cycle? **A**

   A. 0.0001
   B. 0.0005
   C. 0.05
   D. 0.001

5. Small targets are best detected by: **C**

   A. Short pulses transmitted at a fast rate.
   B. Using J band frequencies.
   C. Using a long pulse width with high output power.
   D. All of these answers are correct.

6. What is the relationship between pulse repetition rate and pulse width? **D**

   A. Higher PRR with wider pulse width.
   B. The pulse repetition rate does not change with the pulse width.
   C. The pulse width does not change with the pulse repetition rate.
   D. Lower PRR with wider pulse width.

Key Topic 7 - Components-1

1. What component of a RADAR receiver is represented by block 46 in Fig. 8A1? **B**

A. The ATR box
B. The TR box.
C. The RF Attenuator.
D. The Crystal Detector.

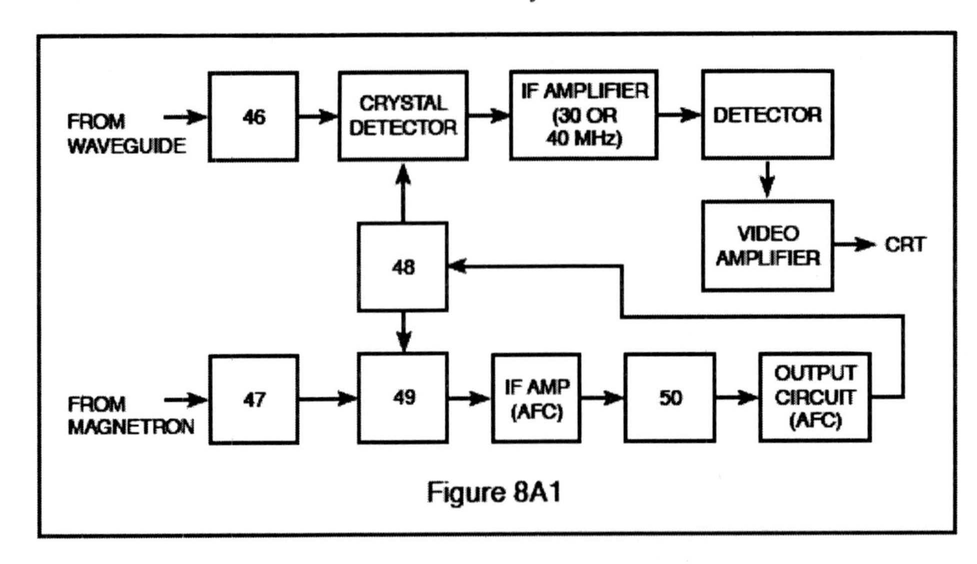

Figure 8A1

2. What component of a RADAR receiver is represented by block 47 in Fig. 8A1? **C**

A. The ATR box.
B. The TR box.
C. The RF Attenuator.
D. The Crystal Detector.

3. When comparing a TTL and a CMOS NAND gate: **A**

A. Both have active pull-up characteristics.
B. Both have three output states.
C. Both have comparable input power sourcing.
D. Both employ Schmitt diodes for increased speed capabilities.

4. Silicon crystals: **D**

A. Are very sensitive to static electric charges.
B. Should be wrapped in lead foil for storage.
C. Tolerate very low currents.
D. All of these.

5. Which is typical current for a silicon crystal used in a RADAR mixer or detector circuit? **A**

A. 3 mA
B. 15 mA
C. 50 mA
D. 100 mA

6. A basic sample-and-hold circuit contains: **D**

A. An analog switch and an amplifier.
B. An analog switch, a capacitor, and an amplifier.
C. An analog multiplexer and a capacitor.
D. An analog switch, a capacitor, amplifiers and input and output buffers.

Key Topic 8 - Components-2

1. The basic frequency determining element in a Gunn oscillator is: **C**

   A. The power supply voltage.
   B. The type of semiconductor used.
   C. The resonant cavity.
   D. The loading of the oscillator by the mixer.

2. Which of the following is not a method of analog-to-digital conversion? **B**

   A. Delta-sigma conversion.
   B. Dynamic-range conversion.
   C. Switched-capacitor conversion.
   D. Dual-slope integration.

3. When comparing TTL and CMOS logic families, which of the following is true: **C**

   A. CMOS logic requires a supply voltage of 5 volts ±20%, whereas TTL logic requires 5 volts ±5%.
   B. Unused inputs should be tied high or low as necessary especially in the CMOS family.
   C. At higher operating frequencies, CMOS circuits consume almost as much power as TTL circuits.
   D. When a CMOS input is held low, it sources current into whatever it drives.

4. The primary operating frequency of a reflex klystron is controlled by the: **A**

   A. Dimensions of the resonant cavity.
   B. Level of voltage on the control grid.
   C. Voltage applied to the cavity grids.
   D. Voltage applied to the repeller plate.

5. A Gunn diode oscillator takes advantage of what effect? **D**

   A. Negative resistance.
   B. Avalanche transit time.
   C. Bulk-effect.
   D. Negative resistance and bulk-effect.

6. Fine adjustments of a reflex klystron are accomplished by: **B**

   A. Adjusting the flexible wall of the cavity.
   B. Varying the repeller voltage.
   C. Adjusting the AFC control system.
   D. Varying the cavity grid potential.

Key Topic 9 - Circuits-1

1. Blocking oscillators operate on the formula of: **A**

   A. T = R x C.
   B. I = E/R.
   C. By using the receiver's AGC.
   D. None of the above are correct.

2. The block diagram of a typical RADAR system microprocessor is shown in Fig. 8A2. Choose the most correct statement regarding this system. **B**

   A. The ALU is used for address decoding.
   B. The Memory and I/O communicate with peripherals.
   C. The control unit executes arithmetic manipulations.
   D. The internal bus is used simultaneously by all units.

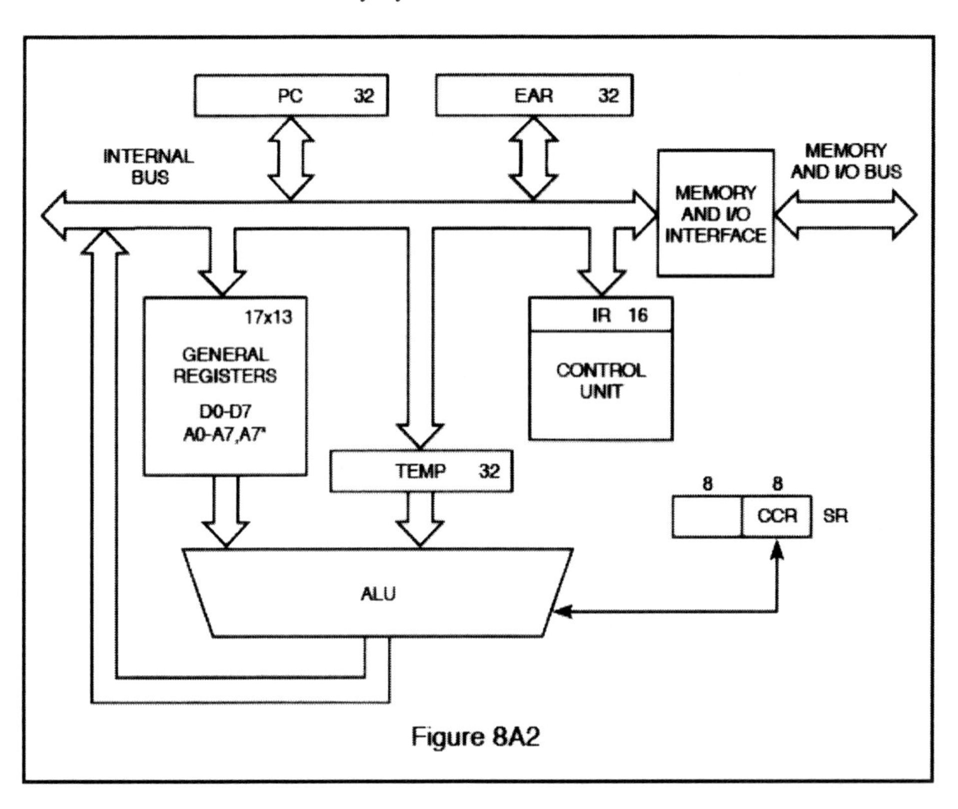

Figure 8A2

3. The phantastron circuit is capable of: **D**

   A. Stabilizing the magnetron.
   B. Preventing saturation of the RADAR receiver.
   C. Being used to control repeller voltage in the AFC system.
   D. Developing a linear ramp voltage when triggered by an external source.

4. The block diagram of a typical RADAR system microprocessor is shown in Fig. 8A2. Choose the most correct statement regarding this system. **A**

   A. The ALU executes arithmetic manipulations.
   B. The ALU is used for address decoding.
   C. General registers are used for arithmetic manipulations.
   D. Address pointers are contained in the control unit.

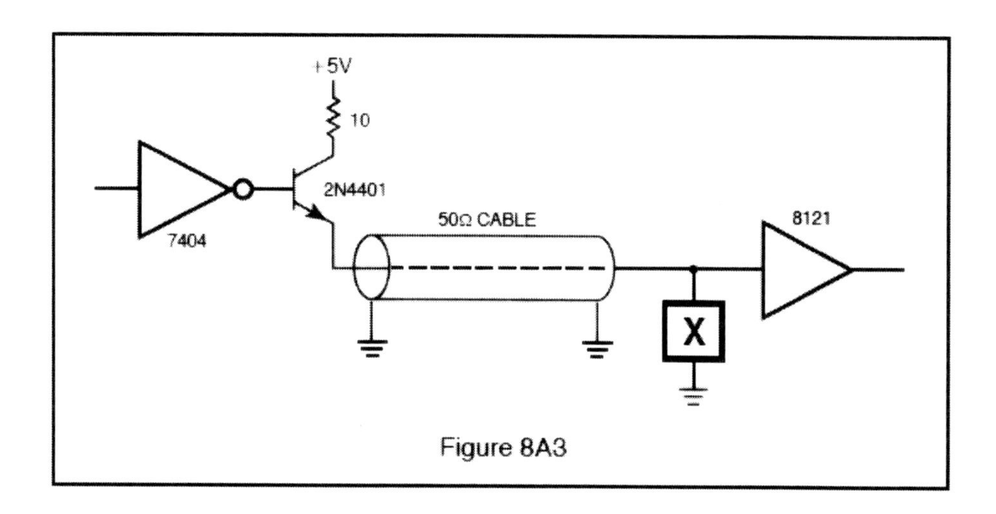

Figure 8A3

5. In the Line-Driver/Coax/Line-receiver circuit shown in Fig. 8A3, what component is represented by the blank box marked "X"? **B**

   A. 25-ohm resistor.
   B. 51-ohm resistor.
   C. 10-microhm inductor.
   D. 20-microhm inductor.

6. Choose the most correct statement: **C**

   A. The magnetron anode is a low voltage circuit.
   B. The anode of the magnetron carries high voltage.
   C. The filament of the magnetron carries dangerous voltages.
   D. The magnetron filament is a low voltage circuit.

Key Topic 10 - Circuits-2

1. In the circuit shown in Fig. 8A4, U5 pins 1 and 4 are high and both are in the reset state. Assume one clock cycle occurs of Clk A followed by one cycle of Clk B. What are the output states of the two D-type flip flops? **D**

   A. Pin 5 low, Pin 9 low.
   B. Pin 5 high, Pin 9 low.
   C. Pin 5 low, Pin 9 high.
   D. Pin 5 high, Pin 9 high.

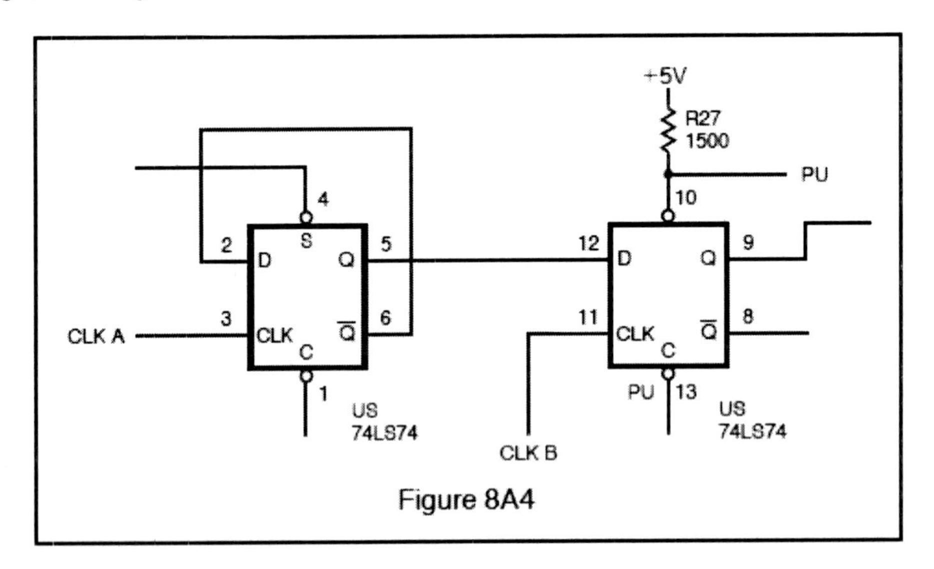

Figure 8A4

2. If more light strikes the photodiode in Fig. 8A5, there will be: **C**

   A. Less diode current.
   B. No change in diode current.
   C. More diode current.
   D. There is wrong polarity on the diode.

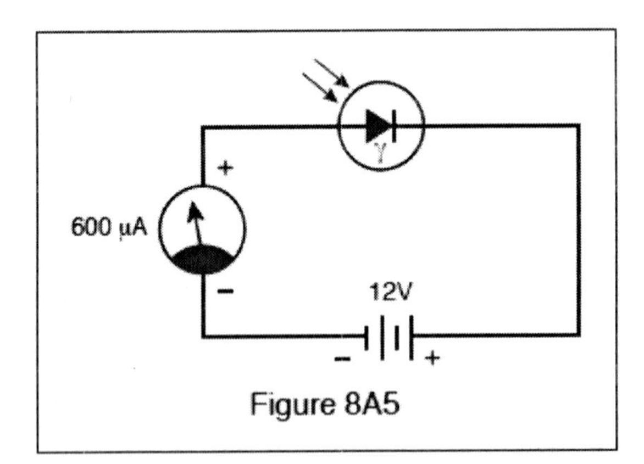

Figure 8A5

3. In the circuit shown in Fig. 8A6, which of the following is true? **B**

A. With A and B high, Q1 is saturated and Q2 is off.
B. With either A or B low, Q1 is saturated and Q2 is off.
C. With A and B low, Q2 is on and Q4 is off.
D. With either A or B low, Q1 is off and Q2 is on.

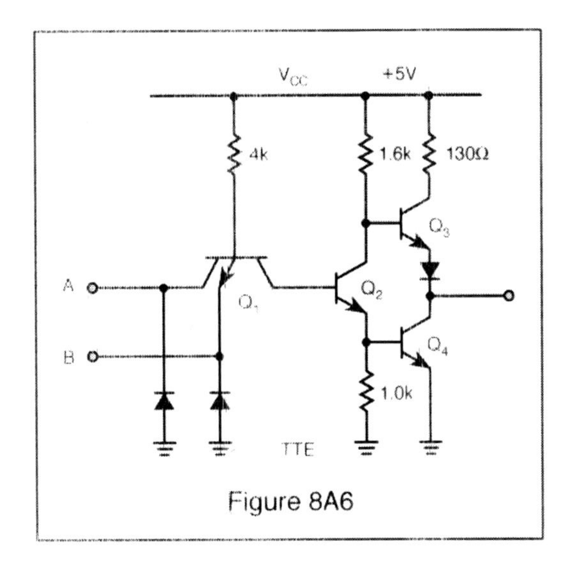

Figure 8A6

4. What is the correct value of RS in Fig. 8A7, if the voltage across the LED is 1.9 Volts with 5 Volts applied and If max equals 40 milliamps? **C**

A. 4,700 ohms.
B. 155 ohms.
C. 77 ohms.
D. 10,000 ohms.

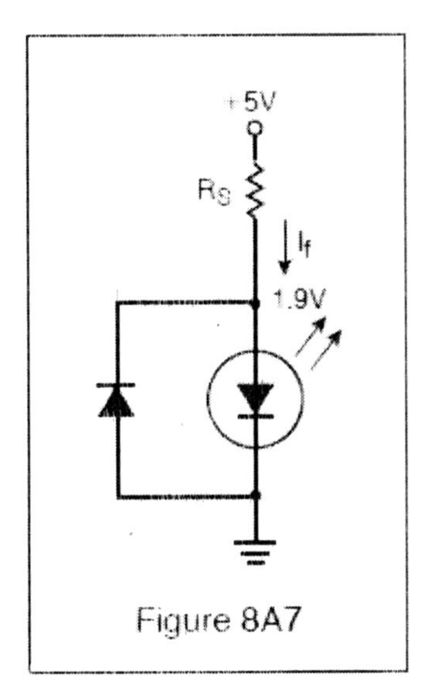

Figure 8A7

5. The block diagram of a typical RADAR system microprocessor is shown in Fig. 8A2. Choose the most correct statement regarding this system. **D**

A. The ALU is used for address decoding.
B. General registers are used for arithmetic manipulations.
C. The control unit executes arithmetic manipulations.
D. Address pointers are contained in the general registers.

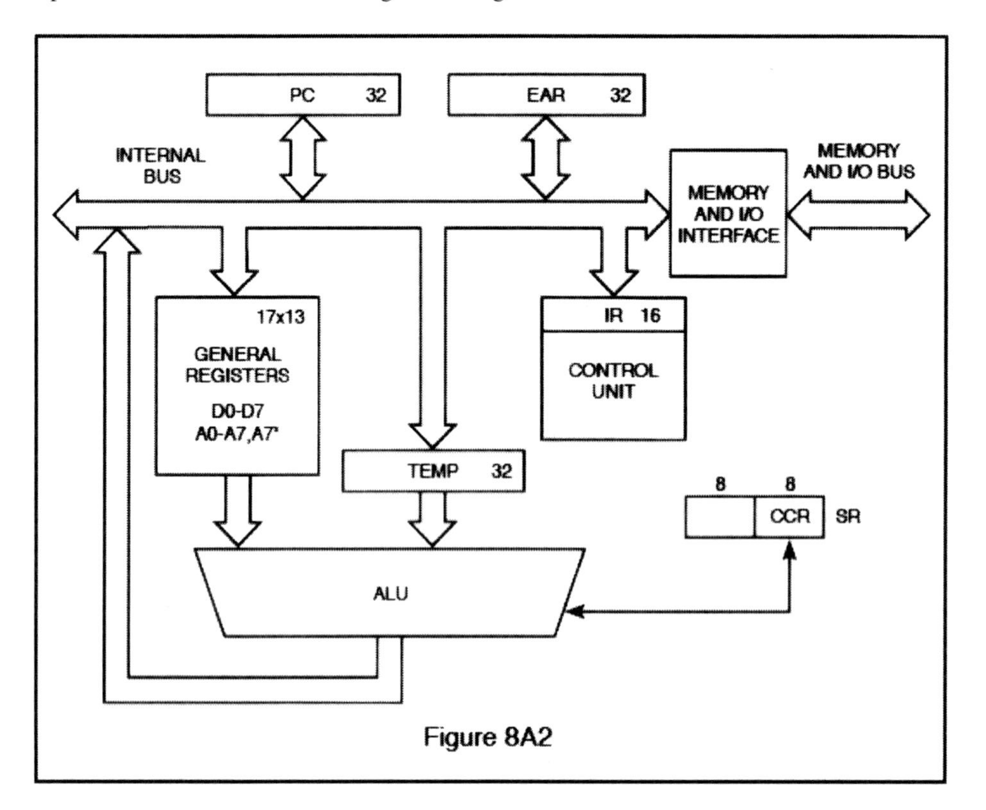

Figure 8A2

6. You are troubleshooting a component on a printed circuit board in a RADAR system while referencing the Truth Table in Fig. 8A8. What kind of integrated circuit is the component? **A**

A. D-type Flip-Flop, 3-State, Inverting.
B. Q-type Flip-Flop, Non-Inverting.
C. Q-type Directional Shift Register, Dual.
D. D to Q Convertor, 2-State.

TRUTH TABLE

INPUTS			OUTPUT
$\overline{OE}$	CP	Dn	$\overline{Qn}$
L	⟋	H	L
L	⟋	L	H
L	L	X	No Change
H	X	X	Z

Note: X = Don't care
Z = High impedance state
⟋ = Low-to-High transition

Figure 8A8

Subelement B - Transmitting Systems: 8 Key Topics - 8 Exam Questions

Key Topic 11 - Transmitting Systems

1. The magnetron is used to: **A**
   - A. Generate the output signal at the proper operating frequency.
   - B. Determine the shape and width of the transmitted pulses.
   - C. Modulate the pulse signal.
   - D. Determine the pulse repetition rate.

2. The purpose of the modulator is to: **B**
   - A. Transmit the high voltage pulses to the antenna.
   - B. Provide high voltage pulses of the proper shape and width to the magnetron.
   - C. Adjust the pulse repetition rate.
   - D. Tune the Magnetron to the proper frequency.

3. Which of the following statements about most modern RADAR transmitter power supplies is false? **A**
   - A. High voltage supplies may produce voltages in excess of 5,000 volts AC.
   - B. There are usually separate low voltage and high voltage supplies.
   - C. Low voltage supplies use switching circuits to deliver multiple voltages.
   - D. Low voltage supplies may supply both AC and DC voltages.

4. The purpose of the Pulse Forming Network is to: **C**
   - A. Act as a low pass filter.
   - B. Act as a high pass filter.
   - C. Produce a pulse of the correct width.
   - D. Regulate the pulse repetition rate.

5. The purpose of the Synchronizer is to: **B**
   - A. Generate the modulating pulse to the magnetron.
   - B. Generate a timing signal that establishes the pulse repetition rate.
   - C. Insure that the TR tube conducts at the proper time.
   - D. Control the pulse width.

6. Which of the following is not part of the transmitting system? **D**
   - A. Magnetron.
   - B. Modulator.
   - C. Pulse Forming Network.
   - D. Klystron.

Key Topic 12 - Magnetrons

1. High voltage is applied to what element of the magnetron? **D**

   A. The waveguide.
   B. The anode.
   C. The plate cap.
   D. The cathode.

2. The characteristic of the magnetron output pulse that relates to accurate range measurement is its: **C**

   A. Amplitude.
   B. Decay time.
   C. Rise time.
   D. Duration.

3. What device is used as a transmitter in a marine RADAR system? **A**

   A. Magnetron.
   B. Klystron.
   C. Beam-powered pentode.
   D. Thyratron.

4. The magnetron is: **D**

   A. A type of diode that requires an internal magnetic field.
   B. A triode that requires an external magnetic field.
   C. Used as the local oscillator in the RADAR unit.
   D. A type of diode that requires an external magnetic field.

5. A negative voltage is commonly applied to the magnetron cathode rather than a positive voltage to the magnetron anode because: **C**

   A. The cathode must be made neutral to force electrons into the drift area.
   B. A positive voltage would tend to nullify or weaken the magnetic field.
   C. The anode can be operated at ground potential for safety reasons.
   D. The cavities might not be shock-excited into oscillation by a positive voltage.

6. The anode of a magnetron is normally maintained at ground potential: **B**

   A. Because it operates more efficiently that way.
   B. For safety purposes.
   C. Never. It must be highly positive to attract the electrons.
   D. Because greater peak-power ratings can be achieved.

Key Topic 13 - Modulation

1. In a solid-state RADAR modulator, the duration of the transmitted pulse is determined by: **C**

   A. The thyratron.
   B. The magnetron voltage.
   C. The pulse forming network.
   D. The trigger pulse.

2. The modulation frequency of most RADAR systems is between: **A**

   A. 60 and 500 Hz.
   B. 3000 and 6000 Hz.
   C. 1500 and 7500 Hz.
   D. 1000 and 3000 Hz.

3. A shipboard RADAR uses a PFN driving a magnetron cathode through a step-up transformer. This results in which type of modulation? **D**

   A. Frequency modulation.
   B. Amplitude modulation.
   C. Continuous Wave (CW) modulation.
   D. Pulse modulation.

4. In a pulse modulated magnetron what device determines the shape and width of the pulse? **A**

   A. Pulse Forming Network.
   B. Thyratron.
   C. LC parallel circuit.
   D. Dimensions of the magnetron cavity.

5. What device(s) may act as the modulator of a RADAR system? **D**

   A. Magnetron.
   B. Klystron.
   C. Video amplifier.
   D. Thyratron or a silicon-controlled rectifier (SCR).

6. The purpose of a modulator in the transmitter section of a RADAR is to: **B**

   A. Improve bearing resolution.
   B. Provide the correct waveform to the transmitter.
   C. Prevent sea return.
   D. Control magnetron power output.

Key Topic 14 - Pulse Forming Networks Modulation

1. The pulse developed by the modulator may have an amplitude greater than the supplyvoltage. This is possible by: **B**
   - A. Using a voltage multiplier circuit.
   - B. Employing a resonant charging choke.
   - C. Discharging a capacitor through an inductor.
   - D. Discharging two capacitors in series and combining their charges.

2. Pulse transformers and pulse-forming networks are commonly used to shape the microwave energy burst RADAR transmitter. The switching devices most often used in such pulse-forming circuits are: **D**
   - A. Power MOSFETS and Triacs.
   - B. Switching transistors.
   - C. Thyratrons and BJT's.
   - D. SCR's and Thyratrons.

3. The purpose of the pulse-forming network is to: **A**
   - A. Determine the width of the modulating pulses.
   - B. Determine the pulse repetition rate.
   - C. Act as a high pass filter.
   - D. Act as a log pass filter.

4. The shape and duration of the high-voltage pulse delivered to the magnetron is established by: **C**
   - A. An RC network in the keyer stage.
   - B. The duration of the modulator input trigger.
   - C. An artificial delay line.
   - D. The time required to saturate the pulse transformer.

5. Pulse-forming networks are usually composed of the following: **B**
   - A. Series capacitors and shunt inductors.
   - B. Series inductors and shunt capacitors.
   - C. Resonant circuit with an inductor and capacitor.
   - D, None of the above.

6. An artificial transmission line is used for: **C**
   - A. The transmission of RADAR pulses.
   - B. Testing the RADAR unit, when actual targets are not available.
   - C. Determining the shape and duration of pulses.
   - D. Testing the delay time for artificial targets.

Key Topic 15 - TR - ATR - Circulators - Directional Couplers-1

1 The ferrite material in a circulator is used as a(an): **D**

A. Electric switch.
B. Saturated reactor.
C. Loading element.
D. Phase shifter.

2. In a circular resonant cavity with flat ends, the E-field and the H-field form with specific relationships. The: **B**

A. E-lines are parallel to the top and bottom walls.
B. E-lines are perpendicular to the end walls.
C. H-lines are perpendicular to the side walls.
D. H-lines are circular to the end walls.

3. A ferrite circulator is most commonly used in what portion of a RADAR system? **C**

A. The antenna.
B. The modulator.
C. The duplexer.
D. The receiver.

4. A circulator provides what function in the RF section of a RADAR system? **A**

A. It replaces the TR cell and functions as a duplexer.
B. It cools the magnetron by forcing a flow of circulating air.
C. It permits tests to be made to the thyristors while in use.
D. It transmits antenna position to the indicator during operation.

5. A directional coupler has an attenuation of -30 db. A measurement of 100 milliwatts at the coupler indicates the power of the line is: **B**

A. 10 watts.
B. 100 watts.
C. 1,000 watts.
D. 10,000 watts.

6. What is the purpose or function of the RADAR duplexer/circulator? **A**

A. An electronic switch that allows the use of one antenna for both transmission and reception.
B. A coupling device that is used in the transition from a rectangular waveguide to a circular waveguide.
C. A modified length of waveguide used to sample a portion of the transmitted energy for testing purposes.
D. A dual section coupling device that allows the use of a magnetron as a transmitter.

Key Topic 16 - TR - ATR - Circulators - Directional Couplers-2

1. The ATR box: **B**

   A. Protects the receiver from strong RADAR signals.
   B. Prevents the received signal from entering the transmitter.
   C. Turns off the receiver when the transmitter is on.
   D. All of the above.

2. When a pulse RADAR is radiating, which elements in the TR box are energized? **C**

   A. The TR tube only.
   B. The ATR tube only.
   C. Both the TR and ATR tubes.
   D. Neither the TR nor ATR tubes.

3. The TR box: **D**

   A. Prevents the received signal from entering the transmitter.
   B. Protects the receiver from the strong RADAR pulses.
   C. Turns off the receiver when the transmitter is on.
   D. Protects the receiver from the strong RADAR pulses and mutes the receiver when the transmitter is on.

4. What device is located between the magnetron and the mixer and prevents received signals from entering the magnetron? **A**

   A. The ATR tube.
   B. The TR tube.
   C. The RF Attenuator.
   D. A resonant cavity.

5. A keep-alive voltage is applied to: **C**

   A. The crystal detector.
   B. The ATR tube.
   C. The TR tube.
   D. The magnetron.

6. A DC keep-alive potential: **D**

   A. Is applied to a TR tube to make it more sensitive.
   B. Partially ionizes the gas in a TR tube, making it very sensitive to transmitter pulses.
   C. Fully ionizes the gas in a TR tube.
   D. Is applied to a TR tube to make it more sensitive and partially ionizes the gas in a TR tube.

Key Topic 17 - Timer - Trigger - Synchronizer Circuits

1. What RADAR circuit determines the pulse repetition rate (PRR)? **B**

   A. Discriminator.
   B. Timer (synchronizer circuit).
   C. Artificial transmission line.
   D. Pulse-rate-indicator circuit.

2. The triggering section is also known as the: **D**

   A. PFN.
   B. Timer circuit.
   C. Blocking oscillator.
   D. Synchronizer.

3. Operation of any RADAR system begins in the: **A**

   A. Triggering section.
   B. Magnetron.
   C. AFC.
   D. PFN.

4. The timer circuit: **D**

   A. Determines the pulse repetition rate (PRR).
   B. Determines range markers.
   C. Provides blanking and unblanking signals for the CRT.
   D. All of the above

5. Pulse RADARs require precise timing for their operation. Which type circuit below might best be used to provide these accurate timing pulses? **A**

   A. Single-swing blocking oscillator.
   B. AFC controlled sinewave oscillator.
   C. Non-symmetrical astable multivibrator.
   D. Triggered flip-flop type multivibrator.

6. Unblanking pulses are produced by the timer circuit. Where are they sent? **C**

   A. IF amplifiers.
   B. Mixer.
   C. CRT.
   D. Discriminator.

Key Topic 18 - Power Supplies

1. An advantage of resonant charging is that it: C

   A. Eliminates the need for a reverse current diode.
   B. Guarantees perfectly square output pulses.
   C. Reduces the high-voltage power supply requirements.
   D. Maintains a constant magnetron output frequency.

2. The characteristics of a field-effect transistor (FET) used in a modern RADAR switching power supply can be compared as follows: **B**

   A. "On" state compares to a bipolar transistor. "Off" state compares to a 1-Megohm resistor.
   B. "On" state compares to a pure resistor. "Off" state compares to a mechanical relay.
   C. "On" state compares to an low resistance inductor. "Off" state compares to a 10-Megohm resistor.
   D. "On" state compares to a resistor. "Off" state compares to a capacitor.

3. A pulse-width modulator in a switching power supply is used to: **C**

   A. Provide the reference voltage for the regulator.
   B. Vary the frequency of the switching regulator to control the output voltage.
   C. Vary the duty cycle of the regulator switch to control the output voltage.
   D. Compare the reference voltage with the output voltage sample and produce an error voltage.

4. In a fixed-frequency switching power supply, the pulse width of the switching circuit will increase when: **A**

   A. The load impedance decreases.
   B. The load current decreases.
   C. The output voltage increases.
   D. The input voltage increases.

5. A major consideration for the use of a switching regulator power supply over a linear regulator is: **D**

   A. The switching regulator has better regulation.
   B. The linear regulator does not require a transformer to step down AC line voltages to a usable level.
   C. The switching regulator can be used in nearly all applications requiring regulated voltage.
   D. The overall efficiency of a switching regulator is much higher than a linear power supply.

6. Which of the following characteristics are true of a power MOSFET used in a RADAR switching supply? **B**

   A. Low input impedance; failure mode can be gate punch-through.
   B. High input impedance; failure mode can be gate punch-through.
   C. High input impedance; failure mode can be thermal runaway.
   D. Low input impedance; failure mode can be gate breakdown.

Subelement C - Receiving Systems: 10 Key Topics - 10 Exam Questions

Key Topic 19 - Receiving Systems

1. Which of the following statements is true? **A**

   A. The front end of the receiver does not provide any amplification to the RADAR signal.
   B. The mixer provides a gain of at least 6 db.
   C. The I.F. amplifier is always a high gain, narrow bandwidth amplifier.
   D. None of the above.

2. Logarithmic receivers: **B**

   A. Can't be damaged.
   B. Can't be saturated.
   C. Should not be used in RADAR systems.
   D. Have low sensitivity.

3. RADAR receivers are similar to: **D**

   A. FM receivers.
   B. HF receivers.
   C. T.V. receivers.
   D. Microwave receivers.

4. What section of the receiving system sends signals to the display system? **A**

   A. Video amplifier.
   B. Audio amplifier.
   C. I.F. Amplifier.
   D. Resolver.

5. What is the main difference between an analog and a digital receiver? **B**

   A. Special amplification circuitry.
   B. The presence of decision circuitry to distinguish between "on" and "off" signal levels.
   C. An AGC stage is not required in a digital receiver.
   D. Digital receivers produce no distortion.

6. In a RADAR receiver, the RF power amplifier: **C**

   A. Is high gain.
   B. Is low gain.
   C. Does not exist.
   D. Requires wide bandwidth.

Key Topic 20 - Mixers

1. The diagram in Fig. 8C9 shows a simplified RADAR mixer circuit using a crystal diode as the first detector. What is the output of the circuit when no echoes are being received? **D**

   A. 60 MHz CW.
   B. 4095 MHz CW.
   C. 4155 MHz CW.
   D. No output is developed.

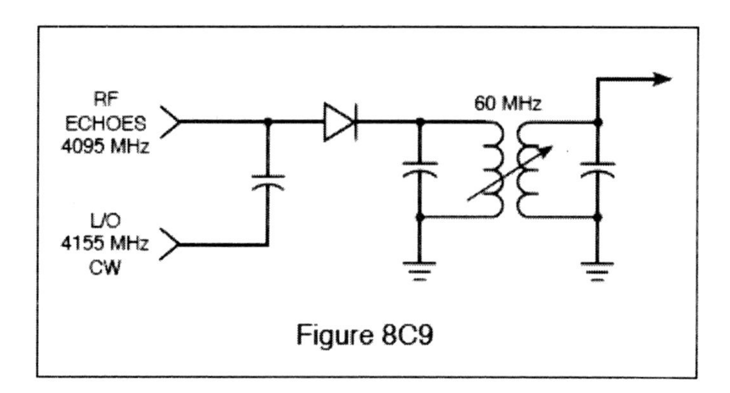

Figure 8C9

2. In the receive mode, frequency conversion is generally accomplished by a: **C**

   A. Tunable wave-guide section.
   B. Pentagrid converter.
   C. Crystal diode.
   D. Ferrite device.

3. An RF mixer has what purpose in a RADAR system? **B**

   A. Mixes the CW transmitter output to form pulsed waves.
   B. Converts a low-level signal to a different frequency.
   C. Prevents microwave oscillations from reaching the antenna.
   D. Combines audio tones with RF to produce the RADAR signal.

4. In a RADAR unit, the mixer uses a: **C**

   A. Pentagrid converter tube.
   B. Field-effect transistor.
   C. Silicon crystal or PIN diode.
   D. Microwave transistor.

5. What component of a RADAR receiver is represented by block 49 in Fig. 8A1? **D**

A. Discriminator.
B. IF amplifier.
C. Klystron.
D. Crystal detector (the mixer).

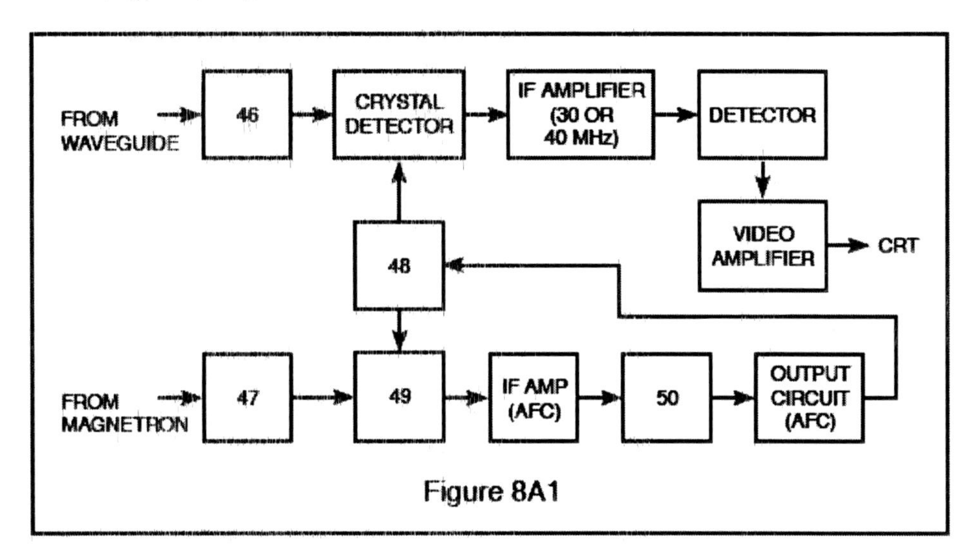

Figure 8A1

6. In a RADAR unit, the mixer uses: **A**

A. PIN diodes and silicon crystals.
B. PIN diodes.
C. Boettcher crystals.
D. Silicon crystals.

Key Topic 21 - Local Oscillators

1. The error voltage from the discriminator is applied to the: **A**

A. Repeller (reflector) of the klystron.
B. Grids of the IF amplifier.
C. Grids of the RF amplifiers.
D. Magnetron.

2. In a RADAR unit, the local oscillator is a: **B**

A. Hydrogen Thyratron.
B. Klystron.
C. Pentagrid converter tube.
D. Reactance tube modulator.

3. What component of a RADAR receiver is represented by block 48 in Fig. 8A1? **A**

A. Klystron (local oscillator).
B. Discriminator.
C. IF amplifier.
D. Crystal detector.

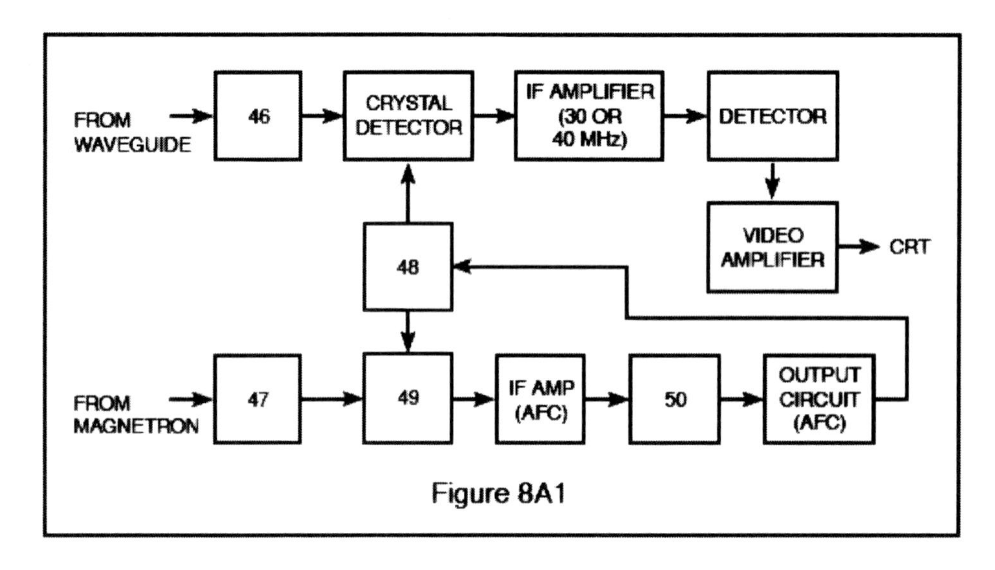

Figure 8A1

4. What device(s) could be used as the local oscillator in a RADAR receiver? **C**

A. Thyratron.
B. Klystron.
C. Klystron and a Gunn Diode.
D. Gunn diode.

5. The klystron local oscillator is constantly kept on frequency by: **B**

A. Constant manual adjustments.
B. The Automatic Frequency Control circuit.
C. A feedback loop from the crystal detector.
D. A feedback loop from the TR box.

6. How may the frequency of the klystron be varied? **D**

A. Small changes can be made by adjusting the anode voltage.
B. Large changes can be made by adjusting the frequency.
C. By changing the phasing of the buncher grids
D. Small changes can be made by adjusting the repeller voltage and large changes can be made by adjusting the size of the resonant cavity.

Key Topic 22 - Amplifiers

1. Overcoupling in a RADAR receiver will cause? **D**

   A. Improved target returns.
   B. Increase the range of the IAGC.
   C. Decrease noise.
   D. Oscillations.

2. The usual intermediate frequency of a shipboard RADAR unit is: **C**

   A. 455 kHz.
   B. 10.7 MHz.
   C. 30 or 60 MHz.
   D. 120 MHz.

3. The I.F. Amplifier bandwidth is: **A**

   A. Wide for short ranges and narrow for long ranges.
   B. Wide for long ranges and narrow for short ranges.
   C. Constant for all ranges.
   D. Adjustable from the control panel.

4. A logarithmic IF amplifier is preferable to a linear IF amplifier in a RADAR receiver because it: **D**

   A. Has higher gain.
   B. Is more easily aligned.
   C. Has a lower noise figure.
   D. Has a greater dynamic range.

5. The high-gain IF amplifiers in a RADAR receiver may amplify a 2 microvolt input signal to an output level of 2 volts. This amount of amplification represents a gain of: **C**

   A. 60 db.
   B. 100 db.
   C. 120 db.
   D. 1,000 db.

6. In a RADAR receiver AGC and IAGC can vary between: **B**

   A. 10 and 15 db.
   B. 20 and 40 db.
   C. 30 and 60 db.
   D. 5 and 30 db.

Key Topic 23 - Detectors - Video Amplifiers

1. Which of the following statements is correct? **C**

   A. The video amplifier is located between the mixer and the I.F. amplifier.
   B. The video amplifier operates between 60 MHz and 120 Mhz.
   C. The video amplifier is located between the I.F. amplifier and the display system.
   D. The video amplifier is located between the local oscillator and the mixer.

2. Video amplifiers in pulse RADAR receivers must have a broad bandwidth because: **A**

   A. Weak pulses must be amplified.
   B. High frequency sine waves must be amplified.
   C. The RADARs operate at PRFs above 100.
   D. The pulses produced are normally too wide for video amplification.

3. In video amplifiers, compensation for the input and output stage capacitances must be accomplished to prevent distorting the video pulses. This compensation is normally accomplished by connecting: **D**

   A. Inductors in parallel with both the input and output capacitances.
   B. Resistances in parallel with both the input and output capacitances.
   C. An inductor in parallel with the input capacitance and an inductor in series with the output capacitance.
   D. An inductor in series with the input capacitance and an inductor in parallel with the output capacitance.

4. Which of the following signals is not usually an input to the video amplifier? **A**

   A. Resolver.
   B. Range.
   C. Brilliance.
   D. Contrast.

5. Which of the following signals are usually an input to the video amplifier? **D**

   A. Range.
   B. Brilliance.
   C. Contrast.
   D. All of the above.

6. The video (second) detector in a pulse modulated RADAR system would most likely use a/an: **B**

   A. Discriminator detector.
   B. Diode detector.
   C. Ratio detector.
   D. Infinite impedance detector.

Key Topic 24 - Automatic Frequency Control - AFC

1. The AFC system is used to: **B**

   A. Control the frequency of the magnetron.
   B. Control the frequency of the klystron.
   C. Control the receiver gain.
   D. Control the frequency of the incoming pulses.

2. A circuit used to develop AFC voltage in a RADAR receiver is called the: **D**

   A. Peak detector.
   B. Crystal mixer.
   C. Second detector.
   D. Discriminator.

3. In the AFC system, the discriminator compares the frequencies of the: **A**

   A. Magnetron and klystron.
   B. PRR generator and magnetron.
   C. Magnetron and crystal detector.
   D. Magnetron and video amplifier.

4. An AFC system keeps the receiver tuned to the transmitted signal by varying the frequency of the: **C**

   A. Magnetron.
   B. IF amplifier stage.
   C. Local oscillator.
   D. Cavity duplexer.

5. A RADAR transmitter is operating on 3.0 GHz and the reflex klystron local oscillator, operating at 3.060 GHz, develops a 60 MHz IF. If the magnetron drifts higher in frequency, AFC system must cause the klystron repeller plate to become: **B**

   A. More positive.
   B. More negative.
   C. Less positive.
   D. Less negative.

6. What component is block 50 in Fig. 8A1? **C**

   A. IF amplifier.
   B. AFC amplifier.
   C. Discriminator.
   D. Crystal detector.

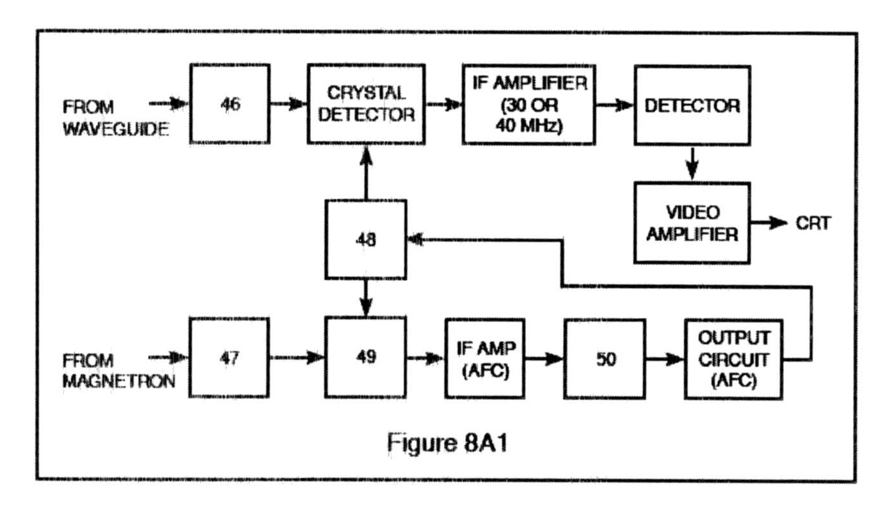

Figure 8A1

Key Topic 25 - Sea Clutter - STC

1. The STC circuit is used to: **D**

   A. Increase receiver stability.
   B. Increase receiver sensitivity.
   C. Increase receiver selectivity.
   D. Decrease sea return on a RADAR receiver.

2. The STC circuit: **B**

   A. Increases the sensitivity of the receiver for close targets.
   B. Decreases sea return on the PPI scope.
   C. Helps to increase the bearing resolution of targets.
   D. Increases sea return on the PPI scope.

3. Sea return is: **C**

   A. Sea water that gets into the antenna system.
   B. The return echo from a target at sea.
   C. The reflection of RADAR signals from nearby waves.
   D. None of the above.

4. Sea clutter on the RADAR scope cannot be effectively reduced using front panel controls. What circuit would you suspect is faulty? **A**

   A. Sensitivity Time Control (STC) circuit.
   B. False Target Eliminator (FTE) circuit.
   C. Fast Time Constant (FTC) circuit.
   D. Intermediate Frequency (IF) circuit.

5. What circuit controls the suppression of sea clutter? **B**

   A. EBL circuit.
   B. STC circuit.
   C. Local oscillator.
   D. Audio amplifier.

6. The sensitivity time control (STC) circuit: **A**

   A. Decreases the sensitivity of the receiver for close objects.
   B. Increases the sensitivity of the receiver for close objects.
   C. Increases the sensitivity of the receiver for distant objects.
   D. Decreases the sensitivity of the transmitter for close objects.

Key Topic 26 - Power Supplies

1. Prior to making "power-on" measurements on a switching power supply, you should be familiar with the supply because of the following: **B**
   A. You need to know where the filter capacitors are so they can be discharged.
   B. If it does not use a line isolation transformer you may destroy the supply with groundedt est equipment.
   C. It is not possible to cause a component failure by using ungrounded test equipment.
   D. So that measurements can be made without referring to the schematic.

2. A constant frequency switching power supply regulator with an input voltage of 165 volts DC, and a switching frequency of 20 kHz, has an "ON" time of 27 microseconds when supplying 1 ampere to its load. What is the output voltage across the load? **C**
   A. It cannot be determined with the information given.
   B. 305.55 volts DC.
   C. 89.1 volts DC.
   D. 165 volts DC.

3. The circuit shown in Fig. 8C10 is the output of a switching power supply. Measuring from the junction of CR6, CR7 and L1 to ground with an oscilloscope, what waveform would you expect to see? **D**
   A. Filtered DC.
   B. Pulsating DC at line frequency.
   C. AC at line frequency.
   D. Pulsating DC much higher than line frequency.

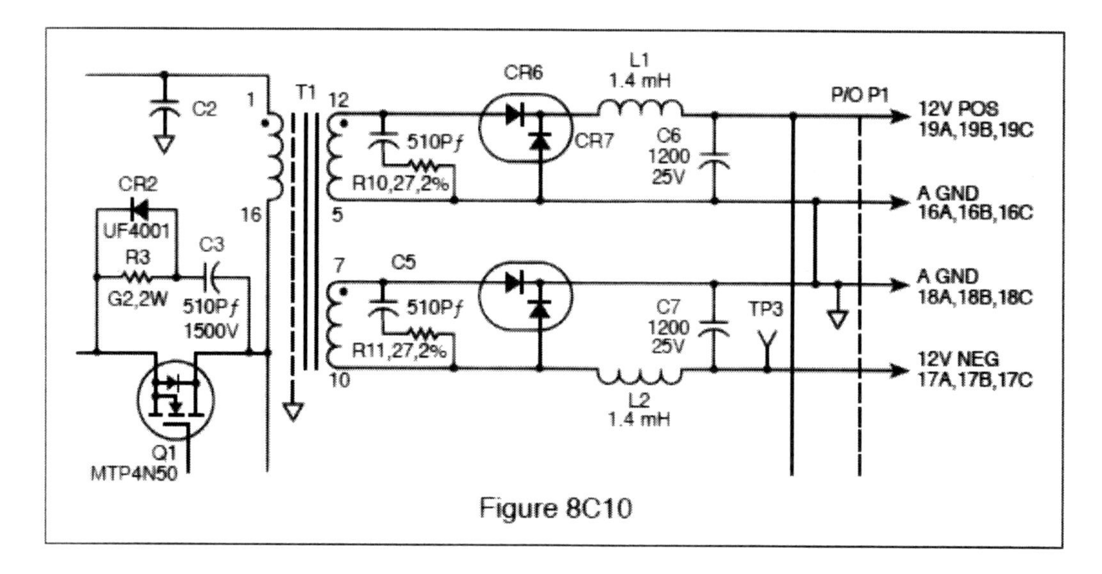

Figure 8C10

4. With regard to the comparator shown in Fig. 8C11, the input is a sinusoid. Nominal high level output of the comparator is 4.5 volts. Choose the most correct statement regarding the input and output. **A**

   A. The leading edge of the output waveform occurs 180 degrees after positive zero crossing of the input waveform.
   B. The rising edge of the output waveform trails the positive zero crossing of the input waveform by 45 degrees.
   C. The rising edge of the output waveform trails the negative zero crossing of the input waveform by 45 degrees.
   D. The rising edge of the output waveform trails the positive peak of the input waveform by 45 degrees.

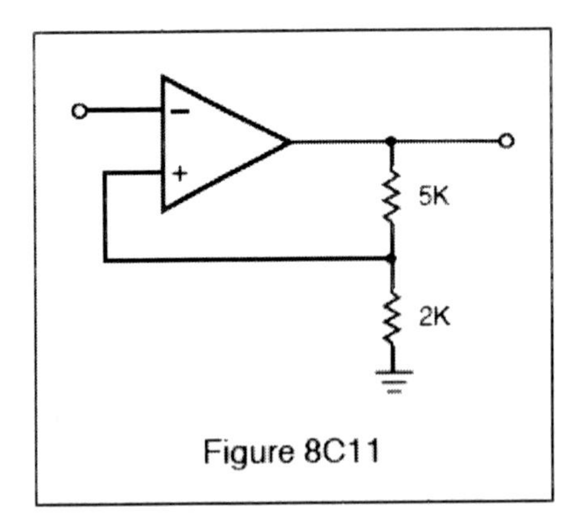

Figure 8C11

5. When monitoring the gate voltage of a power MOSFET in the switching power supply of a modern RADAR, you would expect to see the gate voltage change from "low" to "high" by how much? **C**

   A. 1 volt to 2 volts.
   B. 300 microvolts to 700 microvolts.
   C. Greater than 2 volts.
   D. 1.0 volt to 20.0 volts.

6. The nominal output high of the comparator shown in Fig. 8C11 is 4.5 volts. Choose the most correct statement which describes the trip points. **D**

   A. Upper trip point is 4.5 volts. Lower trip point is approximately 0 volts.
   B. Upper trip point is 2.5 volts. Lower trip point is approximately 2.0 volts.
   C. Upper trip point is 900 microvolts. Lower trip point is approximately 0 volts.
   D. Upper trip point is +1.285 volts. Lower trip point is -1.285 volts.

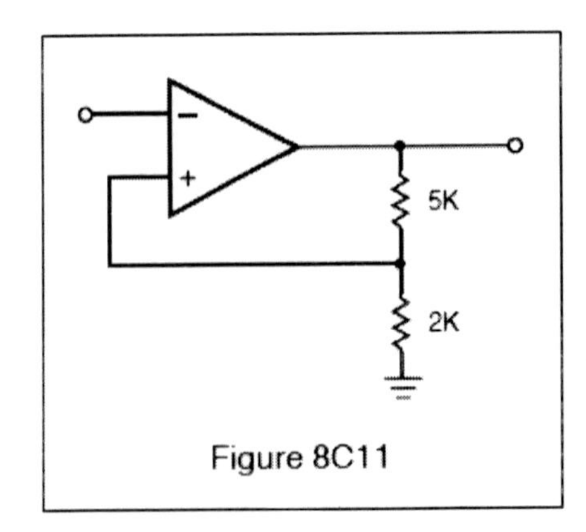

Figure 8C11

Key Topic 27 - Interference Issues

1. One of the best methods of reducing noise in a RADAR receiver is? **B**

   A. Changing the frequency.
   B. Isolation.
   C. Replacing the resonant cavity.
   D. Changing the IF strip.

2. The primary cause of noise in a RADAR receiver can be attributed to: **D**

   A. Electrical causes.
   B. Atmospheric changes.
   C. Poor grounding.
   D. Thermal noise caused by RADAR receiver components.

3. Noise can appear on the LCD as: **A**

   A. Erratic video and sharp changes in intensity.
   B. Black spots on the screen.
   C. Changes in bearings.
   D. None of the above.

4. RADAR interference on a communications receiver appears as: **D**

   A. A varying tone.
   B. Static.
   C. A hissing tone.
   D. A steady tone.

5. In a RADAR receiver the most common types of interference are? **A**

   A. Weather and sea return.
   B. Sea return and thermal.
   C. Weather and electrical.
   D. Jamming and electrical.

6. Noise can: **C**

   A. Mask larger targets.
   B. Change bearings.
   C. Mask small targets.
   D. Increase RADAR transmitter interference.

Key Topic 28 - Miscellaneous

1. The purpose of the discriminator circuit in a RADAR set is to: **C**
   - A. Discriminate against nearby objects.
   - B. Discriminate against two objects with very similar bearings.
   - C. Generate a corrective voltage for controlling the frequency of the klystron local oscillator.
   - D. Demodulate or remove the intelligence from the FM signal.

2. The MTI circuit: **B**
   - A. Acts as a mixer in a RADAR receiver.
   - B. Is a filter, which blocks out stationary targets, allowing only moving targets to be detected.
   - C. Is used to monitor transmitter interference.
   - D. Will pick up targets, which are not in motion.

3. Where is a RF attenuator used in a RADAR unit? **C**
   - A. Between the antenna and the receiver.
   - B. Between the magnetron and the antenna.
   - C. Between the magnetron and the AFC section of the receiver.
   - D. Between the AFC section and the klystron.

4. The condition known as "glint" refers to a shifting of clutter with each RADAR pulse and can be caused by a: **A**
   - A. Improperly functioning MTI filter.
   - B. Memory failure.
   - C Low AFC voltage.
   - D. Interference from electrical equipment.

5. An ion discharge (TR) cell is used to: **D**
   - A. Protect the transmitter from high SWRs.
   - B. Lower the noise figure of the receiver.
   - C. Tune the local oscillator of the RADAR receiver.
   - D. Protect the receiver mixer during the transmit pulse.

6. When the receiver employs an MTI circuit: **B**
   - A. The receiver gain increases with time.
   - B. Only moving targets will be displayed.
   - C. The receiver AGC circuits are disabled.
   - D. Ground clutter will be free of "rabbits."

Subelement D - Display & Control Systems: 10 Key Topics - 10 Exam Questions

Key Topic 29 - Displays

1. Modern liquid crystal displays have a pixel count of: **A**

A. Greater than 200 pixels per inch.
B. Greater than 50 pixels per inch.
C. Can have no more than 125 pixels per inch.
D. Can implement 1,000 pixels per inch.

2. Voltages used in CRT anode circuits are in what range of value? **B**

A. 0.5-10 mV.
B. 10-50 kV.
C. 20-50 mV.
D. 200-1000 V.

3. The purpose of the aquadag coating on the CRT is: **D**

A. To protect the electrons from strong electric fields.
B. To act as a second anode.
C. To attract secondary emissions from the CRT screen.
D. All of the above

4. LCD patterns are formed when: **A**

A. Current passes through the crystal causing them to align.
B. When voltage is reduced to the raster scan display.
C. When the deflection coils are resonant.
D. When the ships antenna's bearing is true North.

5. In a raster-type display, the electron beam is scanned: **B**

A. From the center of the display to the outer edges.
B. Horizontally and vertically across the CRT face.
C. In a rotating pattern which follows the antenna position.
D. From one specified X-Y coordinate to the next.

6. Select the statement, which is most correct regarding a raster scan display. **C**

A. Raster displays are the same as conventional T.V. receivers.
B. The scan rate for a RADAR system is 30 frames per second.
C. Raster scanning is controlled by clock pulses and requires an address bus.
D. Raster scanning is not used in RADAR systems.

Key Topic 30 - Video Amplifiers and Sweep Circuits

1. What are the usual input signals to the video amplifier? **D**

   A. Low level video.
   B. Fixed range rings.
   C. Variable range rings.
   D. All of the above.

2. Which of the following would not normally be an input to the video amplifier? **C**

   A. Fixed range rings.
   B. Variable range rings.
   C. Resolver signal.
   D. Low level video.

3. The purpose of the sweep amplifier is to: **B**

   A. Increase the power of the video amplifier.
   B. Drive the CRT deflection coils.
   C. Drive the resolver coils.
   D. All of the above.

4. How many deflection coils are driven by the sweep amplifier? **C**

   A. 4
   B. 3
   C. 2
   D. 1

5. The main purpose of the sweep generator is to provide: **D**

   A. Antenna information.
   B. Range rings.
   C. Composite video to the cathode of the CRT.
   D. The drive signal to the sweep amplifier.

6. The main purpose of the video amplifier is to provide: **A**

   A. Composite video to the cathode of the CRT.
   B. Resolver signals
   C. Antenna X and Y signals.
   D. Provide the drive signal to the sweep amplifier.

Key Topic 31 - Timing Circuits

1. Timing circuits are used to provide what function? **A**
   - A. Develop synchronizing pulses for the transmitter system.
   - B. Synchronize the antenna and display system.
   - C. Adjust the sea return.
   - D. Control the North Up presentation.

2. The circuit that develops timing signals is called the: **B**
   - A. Resolver.
   - B. Synchronizer.
   - C. Pulse forming network.
   - D. Video amplifier.

3. Which of the following functions is not affected by the timing circuit? **A**
   - A. Resolver output.
   - B. Pulse repetition frequency.
   - C. Sweep drive.
   - D. Modulation.

4. The synchronizer primarily affects the following circuit or function: **C**
   - A. Mixer.
   - B. Receiver.
   - C. Modulator.
   - D. I.F. Amplifier.

5. The output from the synchronizer usually consists of a: **B**
   - A. Sine wave.
   - B. Pulse or square wave.
   - C. Triangle wave.
   - D. None of the above.

6. The sweep drive is initiated by what circuit? **D**
   - A. Resolver.
   - B. Sweep amplifier.
   - C. Video amplifier.
   - D. Synchronizer.

Key Topic 32 - Fixed Range Markers

1. Accurate range markers must be developed using very narrow pulses. A circuit that could be used to provide these high-quality pulses for the CRT is a: **D**
   - A. Ringing oscillator.
   - B. Monostable multivibrator.
   - C. Triggered bi-stable multivibrator.
   - D. Blocking oscillator.

2. Range markers are determined by: **C**
   - A. The CRT.
   - B. The magnetron.
   - C. The timer.
   - D. The video amplifier.

3. A gated LC oscillator, operating at 27 kHz, is being used to develop range markers. If each cycle is converted to a range mark, the range between markers will be: **A**
   - A. 3 nautical miles.
   - B. 6 nautical miles.
   - C. 8 nautical miles.
   - D. 12 nautical miles.

4. What would be the frequency of a range ring marker oscillator generating range rings at 10 nautical miles intervals? **D**
   - A. 24 kHz
   - B. 16 kHz
   - C. 12 kHz
   - D. 8 kHz

5. What is the distance between range markers if the controlling oscillator is operating at 20 kHz? **C**
   - A. 1 nautical miles.
   - B. 2 nautical miles.
   - C. 4 nautical miles.
   - D. 8 nautical miles.

6. What would be the frequency of a range ring marker oscillator generating range rings at intervals of 0.25 nautical miles? **B**
   - A. 161 kHz
   - B. 322 kHz
   - C. 644 kHz
   - D. 1288 kHz

Key Topic 33 - Variable Range Markers

1. The variable range marker signal is normally fed to the input of the: **C**

   A. Sweep amplifier.
   B. Low voltage power supply regulator.
   C. Video amplifier.
   D. Range ring oscillator.

2. The purpose of the variable range marker is to: **A**

   A. Provide an accurate means of determining the range of a moving target.
   B. Provide a bearing line between own ship and a moving target.
   C. Indicate the distance between two different targets.
   D. Provide a means of calibrating the fixed range rings.

3. How is the variable range marker usually adjusted for accuracy? **D**

   A. Adjusting the frequency of the VRM oscillator at the maximum range.
   B. Adjusting the frequency of the VRM oscillator at the minimum range.
   C. Adjusting the readout to match at the median range ring.
   D. The minimum and maximum ranges are aligned with the matching fixed range ring.

4. The panel control for the variable range marker is normally a: **A**

   A. Variable resistor.
   B. Variable inductance.
   C. Variable capacitance.
   D. Variable resolver.

5. An important component of the VRM system is the: **D**

   A. Resolver.
   B. Interference rejection circuit.
   C. STC sensitivity control.
   D. Shift register.

6. Which of the following statements about the Variable Range Marker system is correct? **B**

   A. The VRM is an auxiliary output of the fixed range marker oscillator.
   B. The VRM system develops a single adjustable range ring.
   C. The VRM system is calibrated using a frequency counter.
   D. The VRM system is controlled by a crystal oscillator.

Key Topic 34 - EBL, Azimuth and True Bearing

1. The purpose of the Electronic Bearing Line is to: **B**
   A. Indicate your own vessel's heading.
   B. Measure the bearing of a specific target.
   C. Indicate True North.
   D. Display the range of a specific target.

2. The Electronic Bearing Line is: **D**
   A. The ships heading line.
   B. A line indicating True North.
   C. Used to mark a target to obtain the distance.
   D. A line from your own vessel to a specific target.

3. Which of the following inputs is required to indicate azimuth? **A**
   A. Gyro signals.
   B. Synchronizer
   C. Resolver.
   D. Range rings.

4. Bearing information from the gyro is used to provide the following: **C**
   A. The heading of the nearest target.
   B. Range and bearing to the nearest target.
   C. Vessel's own heading.
   D. The range of a selected target.

5. Which of the following statements about "true bearing" is correct? **B**
   A. The ship's heading flasher is at the top of the screen.
   B. True North is at the top of the screen and the heading flasher indicates the vessel's course.
   C. The true bearing of the nearest target is indicated.
   D. The relative bearing of the nearest target is indicated.

6. A true bearing presentation appears as follows: **C**
   A. The bow of the vessel always points up.
   B. The course of the five closest targets is displayed.
   C. North is at the top of the display and the ship's heading flasher indicates the vessel's course.
   D. The course and distance of the closest target is displayed.

Key Topic 35 - Memory Systems

1. In a digitized RADAR, the 360 degree sweep is divided into how many digitized segments? **D**

   A. 16
   B. 64
   C. 255
   D. 4,096

2. While troubleshooting a memory problem in a raster scan RADAR, you discover that the "REFRESH" cycle is not operating correctly. What type of memory circuit are you working on? **B**

   A. SRAM
   B. DRAM
   C. ROM
   D. PROM

3. The term DRAM stands for: **C**

   A. Digital refresh access memory.
   B. Digital recording access memory.
   C. Dynamic random access memory.
   D. Digital response area motion.

4. How does the dual memory function reduce sea clutter? **A**

   A. Successive sweeps are digitized and compared. Only signals appearing in both sweeps are displayed.
   B. The dual memory system makes the desired targets larger.
   C. It reduces receiver gain for closer signals.
   D. It increases receiver gain for real targets.

5. How many sequential memory cells with target returns are required to display the target? **B**

   A. 1
   B. 2
   C. 4
   D. 8

6. What is the primary purpose of display system memory? **A**

   A. Eliminate fluctuating targets such as sea return.
   B. Display stationary targets.
   C. Display the last available targets prior to a power dropout.
   D. Store target bearings.

Key Topic 36 - ARPA - CAS

1. The ship's speed indication on the ARPA display can be set manually, but does not change with changes in the vessel's speed. What other indication would point to a related equipment failure? **B**
   - A. "GYRO OUT" is displayed on the ARPA indicator.
   - B. "LOG OUT" is displayed on the ARPA indicator.
   - C. "TARGET LOST" is displayed on the ARPA indicator.
   - D. "NORTH UP" is displayed on the ARPA indicator.

2. What does the term ARPA/CAS refer to? **C**
   - A. The basic RADAR system in operation.
   - B. The device which displays the optional U.S.C.G. Acquisition and Search RADAR information on a CRT display.
   - C. The device which acquires and tracks targets that are displayed on the RADAR indicator's CRT.
   - D. The device which allows the ship to automatically steer around potential hazards.

3. Which of the following would not be considered an input to the computer of a collision avoidance system? **D**
   - A. Own ship's exact position from navigation satellite receiver.
   - B. Own ship's gyrocompass heading.
   - C. Own ship's speed from Doppler log.
   - D. Own ship's wind velocity from an anemometer.

4. Which answer best describes a line on the display which indicates a target's position. The speed is shown by the length of the line and the course by the direction of the line. **A**
   - A. Vector.
   - B. Electronic Bearing Line.
   - C. Range Marker.
   - D. Heading Marker.

5. What is the purpose or function of the "Trial Mode" used in most ARPA equipment? **C**
   - A. It selects trial dots for targets' recent past positions.
   - B. It is used to display target position and your own ship's data such as TCPA, CPA, etc.
   - C. It is used to allow results of proposed maneuvers to be assessed.
   - D. None of these.

6. The ARPA term CPA refers to: **D**
   - A. The furthest point a ship or target will get to your own ship's bow.
   - B. Direction of target relative to your own ship's direction.
   - C. The combined detection and processing of targets.
   - D. The closest point a ship or target will approach your own ship.

Key Topic 37 - Display System Power Supplies

1. The display power supply provides the following: **B**
   A. +18 volts DC for the pulse forming network.
   B. 5 volts DC for logic circuits and (12 volts DC for analog and sweep circuits.
   C. 80 volts AC for the antenna resolver circuits.
   D. All of the above.

2. The display power supply provides the following: **D**
   A. 5 volts DC for logic circuits.
   B. ± 12 volts DC for analog and sweep circuits.
   C. 17kV DC for the CRT HV anode.
   D. All of the above.

3. In a display system power supply what is the purpose of the chopper? **A**
   A. It acts as an electronic switch between the raw DC output and the inverter.
   B. It interrupts the AC supply line at a varying rate depending on the load demands.
   C. It regulates the 5 volt DC output.
   D. It pre-regulates the AC input.

4. In a display system power supply, what is the purpose of the inverter? **D**
   A. Inverts the polarity of the DC voltage applied to the voltage regulators.
   B. Provides the dual polarity 12 volt DC supply.
   C. Acts as the voltage regulator for the 5 volt DC supply.
   D. Produces the pulsed DC input voltage to the power transformer.

5. What would be a common switching frequency for a display system power supply? **A**
   A. 18 kHz
   B. 120 Hz
   C. 60 kHz
   D. 120 kHz

6. What display system power supply output would use a tripler circuit? **C**
   A. The logic circuit supply.
   B. The sweep circuit supply.
   C. The HV supply for the CRT anode.
   D. The resolver drive.

Key Topic 38 - Miscellaneous

1. The heading flash is a momentary intensification of the sweep line on the PPI presentation. Its function is to: **C**

   A. Alert the operator when a target is within range.
   B. Alert the operator when shallow water is near.
   C. Inform the operator of the dead-ahead position on the PPI scope.
   D. Inform the operator when the antenna is pointed to the rear of the ship.

2. The major advantage of digitally processing a RADAR signal is: **B**

   A. Digital readouts appear on the RADAR display.
   B. Enhancement of weak target returns.
   C. An improved operator interface.
   D. Rectangular display geometry is far easier to read on the CRT.

3. In order to ensure that a practical filter is able to remove undesired components from the output of an analog-to-digital converter, the sampling frequency should be: **C**

   A. The same as the lowest component of the analog frequency.
   B. Two times the highest component of the analog frequency.
   C. Greater than two times the highest component of the sampled frequency.
   D. The same as the highest component of the sampled frequency.

4. Bearing resolution is: **A**

   A. The ability to distinguish two adjacent targets of equal distance.
   B. The ability to distinguish two targets of different distances.
   C. The ability to distinguish two targets of different elevations.
   D. The ability to distinguish two targets of different size.

5. The output of an RC integrator, when driven by a square wave with a period of much less than one time constant is a: **D**

   A. Sawtooth wave.
   B. Sine wave.
   C. Series of narrow spikes.
   D. Triangle wave.

6. How do you eliminate stationary objects such as trees, buildings, bridges, etc., from the PPI presentation? **B**

   A. Remove the discriminator from the unit.
   B. Use a discriminator as a second detector.
   C. Calibrate the IF circuit.
   D. Calibrate the local oscillator.

Subelement E - Antenna Systems: 5 Key Topics - 5 Exam Questions

Key Topic 39 - Antenna Systems

1. Slotted waveguide arrays, when fed from one end exhibit: **A**

   A. Frequency scan.
   B. High VSWR.
   C. Poor performance in rain.
   D. A narrow elevation beam.

2. A typical shipboard RADAR antenna is a: **B**

   A. Rotary parabolic transducer.
   B. Slotted waveguide array.
   C. Phased planar array.
   D. Dipole.

3. Good bearing resolution largely depends upon: **D**

   A. A high transmitter output reading.
   B. A high duty cycle.
   C. A narrow antenna beam in the vertical plane.
   D. A narrow antenna beam in the horizontal plane.

4. The center of the transmitted lobe from a slotted waveguide array is: **A**

   A. Several degrees offset from a line perpendicular to the antenna.
   B. Perpendicular to the antenna.
   C. Maximum at the right hand end.
   D. Maximum at the left hand end.

5. How does antenna length affect the horizontal beamwidth of the transmitted signal? **B**

   A. The longer the antenna the wider the horizontal beamwidth.
   B. The longer the antenna the narrower the horizontal beamwidth.
   C. The horizontal beamwidth is not affected by the antenna length.
   D. None of the above.

6. What is the most common type of RADAR antenna used aboard commercial maritime vessels? **C**

   A. Parabolic.
   B. Truncated parabolic.
   C. Slotted waveguide array.
   D. Multi-element Yagi array.

Key Topic 40 - Transmission Lines

1. The VSWR of a microwave transmission line device might be measured using: **D**

   A. A dual directional coupler and a power meter.
   B. A network analyzer.
   C. A spectrum analyzer.
   D. A dual directional coupler, a power meter, and a network analyzer.

2. The impedance total (Z_O) of a transmission line can be calculated by $Z_O = \sqrt{L/C}$ when L and C are known. When a section of transmission line contains 250 microhenries of L and 1000 picofarads of C, its impedance total (Z_O) will be: **C**

   A. 50 ohms.
   B. 250 ohms.
   C. 500 ohms.
   D. 1,000 ohms.

3. If long-length transmission lines are not properly shielded and terminated: **B**

   A. The silicon crystals can be damaged.
   B. Communications receiver interference might result.
   C. Overmodulation might result.
   D. Minimal RF loss can result.

4. A certain length of transmission line has a characteristic impedance of 72 ohms. If the line is cut at its center, each half of the transmission line will have a Z_O of: **C**

   A. 36 ohms.
   B. 144 ohms.
   C. 72 ohms.
   D. The exact length must be known to determine Z_O.

5. Standing waves on a transmission line may be an indication that: **D**

   A. All energy is being delivered to the load.
   B. Source and surge impedances are equal to Z_O and Z_L.
   C. The line is terminated in impedance equal to Z_O.
   D. Some of the energy is not absorbed by the load.

6. What precautions should be taken with horizontal waveguide runs? **A**

   A. They should be sloped slightly downwards at the elbow and a small drain hole drilled in the elbow.
   B. They should be absolutely level.
   C. They should not exceed 10 feet in length.
   D. None of the above.

Key Topic 41 - Antenna to Display Interface

1. The position of the PPI scope sweep must indicate the position of the antenna. The sweep and antenna positions are frequently kept in synchronization by the use of: **A**

   A. Synchro systems.
   B. Servo systems.
   C. DC positioning motors.
   D. Differential amplifiers.

2. On a basic synchro system, the angular information is carried on the: **B**

   A. DC feedback signal.
   B. Stator lines.
   C. Deflection coils.
   D. Rotor lines.

3. What is the most common type of antenna position indicating device used in modern RADARs? **A**

   A. Resolvers.
   B. Servo systems.
   C. Synchro transmitters.
   D. Step motors.

4. Which of the following statements about antenna resolvers is correct? **C**

   A. Most resolvers contain a rotor winding and a delta stator winding.
   B. Resolvers consist of a two rotor windings and two stator windings that are 90 degrees apart.
   C. The basic resolver contains a rotor winding and two stator windings that are 90 degrees apart.
   D. Resolvers consist of a "Y" connected rotor winding and a delta connected stator winding.

5. An antenna synchro transmitter is composed of the following: **B**

   A. Three rotor and two stator windings.
   B. Two rotor and three stator windings.
   C. Three rotor and three stator windings.
   D. A single rotor and 3 stator windings.

6. RADAR antenna direction must be sent to the display in all ARPAs or RADAR systems. How is this accomplished? D

   A. 3-phase synchros.
   B. 2-phase resolvers.
   C. Optical encoders.
   D. Any of the above.

Key Topic 42 - Waveguides-1

1. Waveguides can be constructed from: **D**

   A. Brass.
   B. Aluminum.
   C. Copper.
   D. All of the above.

2. A microwave transmission line constructed of a center conductor suspended between parallel conductive ground planes is called: **C**

   A. Microstrip.
   B. Coax.
   C. Stripline.
   D. Waveguide.

3. Waveguide theory is based upon: **A**

   A. The movement of an electromagnetic field.
   B. Current flow through conductive wires.
   C. Inductance.
   D. Resonant charging.

4. A waveguide is used at RADAR microwave frequencies because: **D**

   A. It is easier to install than other feedline types.
   B. It is more rugged than other feedline types.
   C. It is less expensive than other feedline types.
   D. It has lower transmission losses than other feedline types.

5. Waveguide theory is based on the principals of: **C**

   A. Ohm's Law.
   B. High standing waves.
   C. Skin effect and use of 1/4 wave stubs.
   D. None of the above.

6. How is the signal removed from a waveguide or magnetron? **B**

   A. With a thin wire called a T-hook.
   B. With a thin wire called a J-Hook.
   C. With a coaxial connector.
   D. With a waveguide flange joint.

Key Topic 43 - Waveguides-2

1. A rotary joint is used to: **C**

   A. Couple two waveguides together at right angles.
   B. Act as a switch between two waveguide runs.
   C. Connect a stationary waveguide to the antenna array.
   D. Maintain pressurization at the end of the waveguide.

2. Resistive losses in a waveguide are very small because: **A**

   A. The inner surface of the waveguide is large.
   B. The inner surface of the waveguide is small.
   C. The waveguide does not require a ground connection.
   D. The heat remains in the waveguide and cannot dissipate.

3. A right-angle bend in an X-band waveguide must have a radius greater than: **D**

   A. Three inches.
   B. Six inches.
   C. One inch.
   D. Two inches.

4. To insert RF energy into or extract RF energy from a waveguide, which of the following would not be used? **A**

   A. Coupling capacitance.
   B. Current loop.
   C. Aperture window.
   D. Voltage probe.

5. The following is true concerning waveguides: **D**

   A. Conduction is accomplished by the polarization of electromagnetic and electrostatic fields.
   B. Ancillary deflection is employed.
   C. The magnetic field is strongest at the center of the waveguide.
   D. The magnetic field is strongest at the edges of the waveguide.

6. At microwave frequencies, waveguides are used instead of conventional coaxial transmission lines because: **B**

   A. They are smaller and easier to handle.
   B. They have considerably less loss.
   C. They are lighter since they have hollow centers.
   D. Moisture is never a problem with them.

Subelement F - Installation, Maintenance & Repair: 7 Key Topics - 7 Exam Questions

Key Topic 44 - Equipment Faults-1

1. When you examine the RADAR you notice that there is no target video in the center of the CRT. The blank spot gets smaller in diameter as you increase the range scale. What operator front panel control could be misadjusted? **B**
   - A. TUNE.
   - B. Sensitivity Time Control (STC).
   - C. Anti-Clutter Rain (ACR).
   - D. False Target Elimination (FTE).

2. Range rings on the PPI indicator are oval in shape. Which circuit would you suspect is faulty? **D**
   - A. Timing circuit.
   - B. Video amplifier circuit.
   - C. Range marker circuit.
   - D. Sweep generation circuit.

3. What would be the most likely defective area when there is no target video in the center of the CRT and the blank spot gets smaller in diameter as your range scale is increased? **A**
   - A. The TR (TRL) Cell.
   - B. The local oscillator is misadjusted.
   - C. Video amplifier circuit.
   - D. The IF amplifier circuit.

4. While the vessel is docked the presentation of the pier is distorted near the center of the PPI with the pier appearing to bend in a concave fashion. This is a primary indication of what? **C**
   - A. The deflection coils need adjusting.
   - B. The centering magnets at the CRT neck need adjusting.
   - C. The waveguide compensation delay line needs adjusting.
   - D. The CRT filaments are weakening.

5. In a RADAR using digital video processing, a bright, wide ring appears at a fixed distance from the center of the display on all digital ranges. The transmitter is operating normally. What receiver circuit would you suspect is causing the problem? **B**
   - A. VRM circuit.
   - B. Video storage RAM or shift register.
   - C. Range ring generator.
   - D. EBL circuit.

6. The raster scan RADAR display has missing video in a rectangular block on the screen. Where is the most likely problem area? **C**
   - A. Horizontal sweep circuit.
   - B. Power supply.
   - C. Memory area failure.
   - D. Vertical blanking pulse.

Key Topic 45 - Equipment Faults-2

1. A circuit card in a RADAR system has just been replaced with a spare card. You notice the voltage level at point E in Fig. 8F12 is negative 4.75 volts when the inputs are all at 5 volts. The problem is: **D**

   A. The 25 K resistor is open.
   B. The 100 K resistor has been mistakenly replaced with a 50 K resistor.
   C. The op amp is at the rail voltage.
   D. The 50 K resistor has been mistakenly replaced with a 25 K resistor.

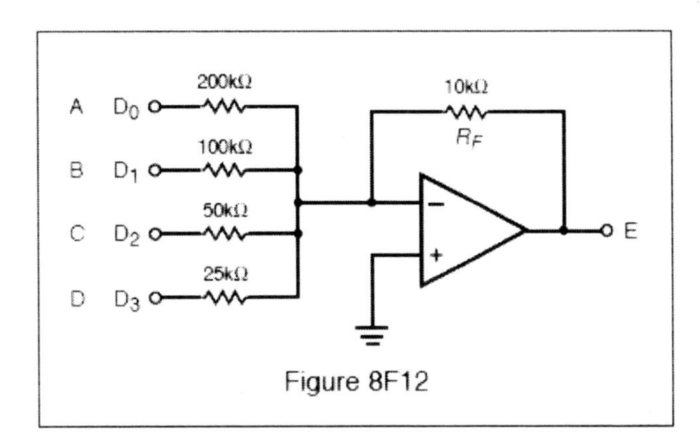

Figure 8F12

2. In the circuit contained in Fig. 8F12, there are 5 volts present at points B and C, and there are zero volts present at points A and D. What is the voltage at point E? **A**

   A. -1.5 Volts.
   B. 3.75 Volts.
   C. 23.75 Volts.
   D. 4.5 Volts.

3. A defective crystal in the AFC section will cause: **B**

   A. No serious problems.
   B. Bright flashing pie sections on the PPI.
   C. Spiking on the PPI.
   D. Vertical spikes that constantly move across the screen.

4. The RADAR display has sectors of solid video (spoking). What would be the first thing to check? **C**

   A. Antenna information circuits failure.
   B. Frequency of raster scan.
   C. For interference from nearby ships.
   D. Constant velocity of antenna rotation.

5. If the TR tube malfunctions: **B**

   A. The transmitter might be damaged.
   B. The receiver might be damaged.
   C. The klystron might be damaged.
   D. Magnetron current will increase.

6. The indicated distance from your own vessel to a lighthouse is found to be in error. What circuit would you suspect? **A**

   A. Range ring oscillator.
   B. Video amplifier.
   C. STC circuit.
   D. FTC circuit.

Key Topic 46 - Equipment Faults-3

1. Silicon crystals are used in RADAR mixer and detector stages. Using an ohmmeter, how might a crystal be checked to determine if it is functional? **B**
   - A. Its resistance should be the same in both directions.
   - B. Its resistance should be low in one direction and high in the opposite direction.
   - C. Its resistance cannot be checked with a dc ohmmeter because the crystal acts as a rectifier.
   - D. It would be more appropriate to use a VTVM and measure the voltage drop across the crystal.

2. In a RADAR unit, if the crystal mixer becomes defective, replace the: **C**
   - A. Crystal only.
   - B. The crystal and the ATR tube.
   - C. The crystal and the TR tube.
   - D. The crystal and the klystron.

3. An increase in magnetron current that coincides with a decrease in power output is an indication of what? **D**
   - A. The pulse length decreasing.
   - B. A high SWR.
   - C. A high magnetron heater voltage.
   - D. The external magnet weakening.

4. It is reported that the RADAR is not receiving small targets. The most likely causes are: **A**
   - A. Magnetron, IF amplifier, or receiver tuning.
   - B. PFN, crystals, or processor memory.
   - C. Crystals, local oscillator tuning, or power supply.
   - D. Fuse blown, IF amp, or video processor.

5. A high magnetron current indicates a/an: **C**
   - A. Defective AFC crystal.
   - B. Increase in duty cycle.
   - C. Defective external magnetic field.
   - D. High standing wave ratio (SWR).

6. Low or no mixer current could be caused by: **D**
   - A. Local oscillator frequency misadjustment.
   - B. TR cell failure.
   - C. Mixer diode degradation.
   - D. All of the above.

Key Topic 47 - Equipment Faults-4

1. If the magnetron is allowed to operate without the magnetic field in place: **B**
   - A. Its output will be somewhat distorted.
   - B. It will quickly destroy itself from excessive current flow.
   - C. Its frequency will change slightly.
   - D. Nothing serious will happen.

2. Targets displayed on the RADAR display are not on the same bearing as their visual bearing. What should you first suspect? **D**
   - A. A bad reed relay in the antenna pedestal.
   - B. A sweep length misadjustment.
   - C. One phase of the yoke assembly is open.
   - D. Incorrect antenna position information.

3. Loss of distant targets during and immediately after wet weather indicates: **A**
   - A. A leak in waveguide or rotary joint.
   - B. High atmospheric absorption.
   - C. Dirt or soot on the rotary joint.
   - D. High humidity in the transmitter causing power supply loading.

4. In a marine RADAR set, a high VSWR is indicated at the magnetron output. The waveguide and rotary joint appear to be functioning properly. What component may be malfunctioning? **D**
   - A. The magnetron
   - B. The waveform generator
   - C. The STC circuit
   - D. The waveguide array termination

5. On a vessel with two RADARs, one has a different range indication on a specific target than the other. How would you determine which RADAR is incorrect? **A**
   - A. Check the sweep and timing circuits of both indicators for correct readings.
   - B. Triangulate target using the GPS and visual bearings.
   - C. Check antenna parallax.
   - D. Use the average of the two indications and adjust both for that amount.

6. An increase in the deflection on the magnetron current meter could likely be caused by: **C**
   - A. Insufficient pulse amplitude from the modulator.
   - B. Too high a B1 level on the magnetron.
   - C. A decrease of the magnetic field strength.
   - D. A lower duty cycle, as from 0.0003 to 0.0002.

Key Topic 48 - Maintenance

1. A thick layer of rust and corrosion on the surface of the parabolic dish will have what effect? **C**

   A. No noticeable effect.
   B. Scatter and absorption of RADAR waves.
   C. Decrease in performance, especially for weak targets.
   D. Slightly out of focus PPI scope.

2. The echo box is used for: **B**

   A. Testing the wavelength of the incoming echo signal.
   B. Testing and tuning of the RADAR unit by providing artificial targets.
   C. Amplification of the echo signal.
   D. Detection of the echo pulses.

3. What should be done to the interior surface of a waveguide in order to minimize signal loss? **C**

   A. Fill it with nitrogen gas.
   B. Paint it with nonconductive paint to prevent rust.
   C. Keep it as clean as possible.
   D. Fill it with a high-grade electrical oil.

4. Which of the following is the most useful instrument for RADAR servicing? **A**

   A. Oscilloscope.
   B. Frequency Counter.
   C. R. F. Wattmeter.
   D. Audio generator.

5. A non-magnetic screwdriver should always be used when replacing what component? **D**

   A. TR tube.
   B. Mixer.
   C. Video amplifier.
   D. Magnetron.

6. What kind of display would indicate water in the waveguide? B

   A. Spoking.
   B. Large circular rings near the center.
   C. Loss of range rings.
   D. Wider than normal targets.

Key Topic 49 - Installation

1. Why is coaxial cable often used for S-band installations instead of a waveguide? **A**
   - A. Losses can be kept reasonable at S-band frequencies and the installation cost is lower.
   - B. A waveguide will not support the power density required for modern S-band RADAR transmitters.
   - C. S-band waveguide flanges show too much leakage and are unsafe for use near personnel.
   - D. Dimensions for S-band waveguide do not permit a rugged enough installation for use by ships at sea.

2. RADAR interference to a communications receiver is eliminated by: **B**
   - A. Not operating other devices when RADAR is in use.
   - B. Properly grounding, bonding, and shielding all units.
   - C. Using a high pass filter on the power line.
   - D. Using a link coupling.

3. Why should long horizontal runs of waveguide be avoided? **D**
   - A. They must be insulated to prevent electric shock.
   - B. To prevent damage from shipboard personnel.
   - C. To minimize reception of horizontally polarized returns.
   - D. To prevent accumulation of condensation.

4. Long horizontal sections of waveguides are not desirable because: **A**
   - A. Moisture can accumulate in the waveguide.
   - B. The waveguide can sag, causing loss of signal.
   - C. Excessive standing waves can occur.
   - D. The polarization of the signal might shift.

5. In a RADAR system, waveguides should be installed: **B**
   - A. Slightly bent for maximum gain.
   - B. As straight as possible to reduce distortion.
   - C. At 90 degree angles to improve resonance.
   - D. As long as possible for system flexibility.

6. What is the most important factor to consider in locating the antenna? **C**
   - A. Allow the shortest cable/waveguide run.
   - B. Maximum height for best long range operation.
   - C. The antenna is in a location that is not shadowed by other structures.
   - D. Easy access for maintenance.

Key Topic 50 - Safety

1. Choose the most correct statement with respect to component damage from electrostatic discharge: **D**
   - A. ESD damage occurs primarily in passive components which are easily identified and replaced.
   - B. ESD damage occurs primarily in active components which are easily identified and replaced.
   - C. The technician will feel a small static shock and recognize that ESD damage has occurred to the circuit.
   - D. ESD damage may cause immediate circuit failures, but may also cause failures much later at times when the RADAR set is critically needed.

2. Before testing a RADAR transmitter, it would be a good idea to: **C**
   - A. Make sure no one is on the deck.
   - B. Make sure the magnetron's magnetic field is far away from the magnetron.
   - C. Make sure there are no explosives or flammable cargo being loaded.
   - D. Make sure the Coast Guard has been notified.

3. While making repairs or adjustments to RADAR units: **B**
   - A. Wear fire-retardant clothing.
   - B. Discharge all high-voltage capacitors to ground.
   - C. Maintain the filament voltage.
   - D. Reduce the magnetron voltage.

4. While removing a CRT from its operating casing, it is a good idea to: **C**
   - A. Discharge the first anode.
   - B. Test the second anode with your fingertip.
   - C. Wear gloves and goggles.
   - D. Set it down on a hard surface.

5. If a CRT is dropped: **D**
   - A. Most likely nothing will happen because they are built with durability in mind.
   - B. It might go out of calibration.
   - C. The phosphor might break loose.
   - D. It might implode, causing damage to workers and equipment.

6. Prior to removing, servicing or making measurements on any solid state circuit boards from the RADAR set, the operator should ensure that: **A**
   - A. The proper work surfaces and ESD grounding straps are in place to prevent damage to the boards from electrostatic discharge.
   - B. The waveguide is detached from the antenna to prevent radiation.
   - C. The magnetic field is present to prevent over-current damage or overheating from occurring in the magnetron.
   - D. Only non-conductive tools and devices are used.

Federal Communications Commission

All Commercial Radio Operator Licenses are issued for the lifetime of the holder by the FCC National Operator Licensing Office:

Federal Communications Commission
1270 Fairfield Road
Gettysburg, PA 17325-7245

The national FCC Information phone number is (888) 225-5322. However, we have found that the following number to be a preferred choice for contacting the FCC with operator license questions or requesting information: (877) 480-3201 (Select Option 2).

For additional sources and information, contact:

FCC Wireless Telecommunications Bureau
Federal Communications Commission
445 12th Street SW
Washington, DC 20554-0004

Phone: (888) 225-5322
Videophone: (844) 432-2275
Fax: (866) 418-0232

FCC License Examinations

The FCC is not involved with administering the actual operator license examination. Many years ago, exams were given at regional FCC government offices, but this was transferred to private, voluntary examiners.

Please refer to page 12 in this manual for the source of national organizations to contact to arrange your examination appointment. Exams are given on a regular basis throughout all states and at some U.S. Military installations worldwide.

Study Aids - References

THE COMPLETE FCC LICENSE HOME-STUDY COURSE:

This is the Original Home-Study "Crash-Course" that includes audio training lessons on CD discs, Training Manuals and course outline Notebook by author Warren Weagant. The Self-Study lessons provide beginners with fundamental concepts and time-tested preparation for passing the federal FCC Exam.

Listening to programmed audio lessons on 8 CDs makes learning faster, easier and more permanent. Audio CD discs permit you to hear and absorb FCC material at any time and at any place you choose. Listen at work, home or car to turn unproductive time into FCC training sessions.

This proven, nationally accepted home study course was created for people who need a better understanding of vital basics and builds question by question for maximum understanding, even for people with no electronics background. Learn everything in the fastest time possible to get your FCC License!

For more information, call for a free brochure: (800) 932-4268 or go to the FCC License Home-Study Course website to read more: www.LicenseTraining.com

RADAR TECHNOLOGY TRAINING GUIDE:

A comprehensive and informative guide for learning more about the principals of Radar Technology and Equipment. This was created for anyone wanting to know more about Radar to be able to pass the FCC Element 8 exam, or to have better knowledge of Radar fundamentals. The format is an "eBook" that is playable on any computer. Simply insert the disc and learn more about Radar Transmitters, Receivers, Subsystems, Indicators, Antennas, Systems and Maintenance. $25

For more information on Command Productions training courses, call (800) 932-4268 for more information to be sent to you, email us at info@commandproductions.com or simply go to our website: www.LicenseTraining.com

Reader Comments & Feedback

Your comments are always appreciated and help us to continue to publish the most up to date and relevant license training study materials.

We would like to hear from you about your experience taking the FCC License examination, your score and how well prepared you were using our study material.

You may call us with your feedback and comments toll free: (800) 932-4268.

Or, send an email to: FCC@CommandProductions.com. Please provide us with your name, address and telephone number.

Warren Weagant
FCC License Training
Command Productions
PO Box 3000
Sausalito, CA 94966-3000

CPSIA information can be obtained
at www.ICGtesting.com
Printed in the USA
FSOW02n1139170118
43385FS